T0212179

INTERNATIONAL CENTRE FOR MECHANICAL SCIENCES

COURSES AND LECTURES No. 227

NONLINEAR DYNAMICS OF ELASTIC BODIES

EDITED BY

Z. WESOLOWSKI

POLISH ACADEMY OF SCIENCES

SPRINGER-VERLAG WIEN GMBH

This work is subject to copyright.
All rights are reserved,
whether the whole or part of the material is concerned
specifically those of translation, reprinting, re-use of illustrations,
broadcasting, reproduction by photocopying machine
or similar means, and storage in data banks.
© 1978 by Springer-Verlag Wien
Originally published by Springer Verlag Wien New York in 1978

ISBN 978-3-211-81512-0 ISBN 978-3-7091-2746-9 (eBook)
DOI 10.1007/978-3-7091-2746-9

PREFACE

This volume contains the texts of five series of lectures devoted to the nonlinear dynamics of elastic bodies which were delivered at the Department of Mechanics of Solids of the International Centre for Mechanical Sciences, Udine, Italy.

The contributions of the various authors are closely interrelated. The two first papers, by T. Manacorda and Cz. Woźniak, provide a basis for the analysis of the problems illustrated by the other lecturers. These include acceleration waves and progression waves in nonlinear elastic materials (Z. Wezolowski) and the stability of elastic systems (S.J. Britvec).

Finally, the contribution by B.R. Seth is of a somewhat different nature. It advocates, for large deformations, the use of generalized measures and discusses the ensuing results and advantages.

We hope the contributions presented will be of interest to research workers involved in investigating the nonlinear response of material s under various static and dynamic conditions.

Z. Wezolowski

LIST OF CONTRIBUTORS

Tristano Manacorda	Università di Pisa, Istituto di Matematiche Applicate, Pisa.
Czeslaw Woźniak	University of Warsaw, Miedzynarodowa 58 m. 63, 03-922 Warszawa, Poland.
Zbigniew Wesolowski	Institute of Fundamental Technological Research, Swietokrzyska 21, Warsaw, Poland.
S.J. Britvec	Professor of Engineering Mechanics. University of Stuttgart and University of Zagreb.
B.R. Seth	Birla Institute of Technology, Mesra, Ranchi, India.

TOPICS IN ELASTODYNAMICS

TRISTANO MANACORDA
Università di Pisa
Istituto di Matematiche Applicate

CHAPTER I
INTRODUCTION. MOTION AND DEFORMATION

1 - DEFORMATIONS.

The notion of a continuous body is a primitive concept. A continuous body can be put in one-to-one correspondence with regions of the Euclidean space; more exactly, with family of such regions. Each of these regions is called a *configuration* of the body.

Let B_0 and B be two different configurations of the same continuous body $\mathcal{B}$, $\underline{X}$ and $\underline{x}$ the positions, in B_0 and B, of the same particle of $\mathcal{B}$ in a fixed frame of reference. The mapping $B_0 \to B$ is a *deformation* of the body. A deformation is *regular* if:

1) the correspondence $\underline{X} \to \underline{x}$ is one-to-one;

2) if we put

$$\underline{x} = \underline{\chi}\ (\underline{X}) \qquad\qquad \underline{X} = \underline{\chi}^{-1}(\underline{x}) \tag{1.1}$$

the functions $\underline{\chi}$ and $\underline{\chi}^{-1}$ are continuous up to their third derivatives (1);

3) the determinant of the *deformation gradient*

$$\underline{F} = \text{Grad}\ \underline{x} = \text{Grad}\ \underline{\chi} = ||x^h{}_{,H}||\ , \quad x^h{}_{,H} = \frac{\partial x^h}{\partial X^H} \tag{1.2}$$

$$h,H = 1,2,3$$

is strictly positive:

$$J = \det \underline{F} > 0.$$

Let $d\underline{x}$ be a linear element of B, and $d\underline{X}$ the corresponding one in the reference configuration B_o; obviously

$$d\underline{x} = \underline{F}\ d\underline{X} \qquad , \qquad d\underline{X} = \underline{F}^{-1} d\underline{x} \quad ^{(2)} \tag{1.3}$$

(1) In these lectures we shall only need the continuity of the second derivatives, but the third derivatives must be considered for the so called congruence conditions.

(2) Let $d\underline{x}^h$ be a triad of linear elements issuing from $\underline{x}$ and non coplanar; the volume v of the thetraedron $d\underline{x}^1$, $d\underline{x}^2$, $d\underline{x}^3$ is given by $v = d\underline{x}^1 \cdot d\underline{x}^2 \times d\underline{x}^3$, and therefore by $v = \underline{F}d\underline{X}^1 \cdot \underline{F}d\underline{X}^2 \times \underline{F}d\underline{X}^3 = J\ d\underline{X}^1 \cdot d\underline{X}^2 \times d\underline{X}^3$, as is well known. Therefore, the condition $J > 0$ is equivalent to the condition that the two triads $d\underline{x}^h$ and $d\underline{X}^H$ be equally oriented.

so that the knowledge of the gradient of deformation implies the knowledge of the deformation of a first order neighbourhood of $\underline{X}$. For instance, from (1.3) it is easily obtained

$$(d\underline{x})^2 = d\underline{X}\cdot\underline{C}\ d\underline{X} \tag{1.4}$$

with

$$\underline{C} = \underline{F}^T\underline{F}. \tag{1.5}$$

$\underline{C}$ is a symmetric tensor defined on B_o, called the right Cauchy-Green deformation tensor.

The deformation tensor $\underline{E}$ is given by

$$\underline{C} = \underline{1} + 2\ \underline{E} \tag{1.6}$$

where $\underline{1}$ is the unit tensor, whose cartesians components δ_{nk} are 0 if $h \neq k$, 1 if $h = k$.

The proper values λ_H of $\underline{C}$ are strictly positive, as it can easily proved using (1.4). Let $\underline{U}$ be a symmetric, positive definite tensor whose proper values coincide with $\sqrt{\lambda_H}$, $H = 1,2,3$; the polar decomposition od $\underline{F}$

$$\underline{F} = \underline{R}\ \underline{U} \tag{1.7}$$

where $\underline{R}$ is a proper rotation, $\underline{R}\ \underline{R}^T = \underline{1}$, det R = 1, is easily proved.

In fact, put

$$\gamma = \underline{F}\ \underline{U}^{-1}; \tag{1.8}$$

we obtain successively

$$\underline{\gamma}^T\underline{\gamma} = (\underline{U}^{-1})^T\underline{F}^T\underline{F}\ \underline{U}^{-1} = \underline{U}^{-1}\underline{C}\ \underline{U}^{-1} = \underline{U}^{-1}\underline{U}^2\ \underline{U}^{-1} = \underline{1}\ ,$$

$$\det.\underline{\gamma} = \det.\underline{F}\ \det.\ (\underline{U}^{-1}) = J\ /\ \det.\underline{U} = J\ /\ |\det.\ \underline{F}| = 1$$

if $J = \det.\underline{F} > 0$. The uniqueness of $\underline{R}$ can be proved by assuming the existence of a differente rotation, say $\underline{R}'$, and proving that this assumption is false.

The symmetric, positive definite tensor

$$\underline{B} = \underline{F}\,\underline{F}^T = \underline{V}^2 \tag{1.9}$$

is called the left Cauchy-Green deformation tensor. $\underline{B}$ is an Eulerian (or local) tensor.

The following identities hold:

$$\underline{B} = \underline{R}\,\underline{C}\,\underline{R}^T \quad , \quad \underline{C} = \underline{R}^T\underline{B}\,\underline{R} . \tag{1.10}$$

$\underline{C}$ is a Lagrangian (or molecular) tensor; his proper directions are called the principal directions of deformation. The rotation $\underline{R}$ brings the proper directions of $\underline{C}$ into coincidence with the proper directions of $\underline{B}$ (3).

$\underline{F}$ determines also the correspondence between oriented surface elements.

Let $d\sigma$ be a surface element of B, whose unit normal is $\underline{n}$, and let $d\Sigma$ be the corresponding element of B_o and $\underline{N}$ the unit normal to $d\Sigma$, whose orientation is choosen so that $\underline{n}\cdot\underline{F}\,\underline{N} > 0$ (in general, $\underline{F}\,\underline{N}$ is neither normal nor tangent to $d\sigma$). Then

$$d\sigma\,\underline{n} = d\Sigma\; J\; (\underline{F}^T)^{-1}\underline{N}. \tag{1.11}$$

Let the reference configuration be changed in B_o', and let the gradient of deformation $B_o' \rightarrow B$ be $\underline{F}'$; obviously,

(3) Occasionally, it will be convenient to introduce the tensor

$$\underline{c} = (\underline{F}^T)^{-1}\underline{F}^{-1} = \underline{B}^{-1} = \underline{1} + 2\underline{e} ;$$

$\underline{c}$ possesses the same proper directions as $\underline{B}$, but his proper numbers are the reciprocal of the proper numbers of $\underline{B}$.

$$\underline{F} = \underline{F}'\,\underline{P}' \quad , \quad \underline{F}' = \underline{F}\,\underline{P} \quad , \tag{1.12}$$

where

$$\underline{F} = \| x^h{}_{,H} \| \quad , \quad \underline{F}' = \| x^h{}_{,H'} \|$$

and

$$\underline{P}' = \| X^H{}_{,H'} \| \tag{1.13}$$

is the gradient of deformation from B_o to B_o'. For instance, we obtain

$$\underline{C}' = \underline{F}'^T\,\underline{F}' = \underline{P}^T \underline{F}^T \underline{F}\,\underline{P} = \underline{P}^T\,\underline{C}\,\underline{P} \ . \tag{1.14}$$

In particular, let us consider the case $\underline{P}' = \underline{Q}$, with $\underline{Q}$ orthogonal; we obtain

$$\underline{C}' \doteq \underline{Q}\,\underline{C}\,\underline{Q}^T \quad , \tag{1.15}$$

and $\underline{C}$ and $\underline{C}'$ have the same principal invariants. If $\underline{P}'$ reduces to a uniform dilatation, $\underline{P}' = d\,\underline{1}$

$$\underline{C}' = d^{-2}\,\underline{C} \tag{1.16}$$

and $\underline{C}$ and $\underline{C}'$ have the same principal directions. More generally, for a conformal transformation

$$\underline{P}' = d\,\underline{R} \quad , \quad (\underline{R}\,\underline{R}^T = \underline{1} \ , \ \det.\ \underline{R} = 1) \tag{1.17}$$

$$\underline{C}' = d^{-2}\,\underline{R}\,\underline{C}\,\underline{R}^T \ . \tag{1.18}$$

Till now we have reviewed some geometrical concepts connected with the deformation of B; we now discuss the motion of B.

The motion of B is an one-to-one mapping of B on one-parameter family of configurations B_t, where the parameter is the time t. In place of (1.1), we can write

$$\underline{x} = \underline{X}_t\ (\underline{X}) = \underline{X}\ (t,\underline{X}) \tag{1.19}$$

which gives the motion of the particle which, at some fixed instant, say t_o, is at $\underline{X}$. For fixed $\underline{X}$, (1.19) is the parametric equation of the trajectory of the particle.

The velocity and the acceleration of a particle are the lagrangean derivatives of $\underline{X}$,

$$\underline{v} = \dot{\underline{x}} = \frac{\partial \underline{X}}{\partial t}\ (^4)\ ,\quad \underline{a} = \ddot{\underline{x}} = \frac{\partial^2 \underline{X}}{\partial t^2} \tag{1.20}$$

We shall suppose $\underline{v}$ and $\underline{a}$ to be continuous for all t in some interval $[t_o, t_1]$.

2 - BALANCE EQUATIONS.

The conservation principles of continuum mechanics, which originate the balance equations, are

1) the principle of conservation of mass;
2) the principle of balance of the linear momentum;
3) the principle of balance of the moment of momentum.

The first principle can be expressed in an integral form by

$$\int_b \rho\, dv = \int_{b_o} \rho_o\, dV \tag{2.1}$$

where b is any subset of B and b_o the corresponding subset of B_o. If the integrands are continuous functions, and if (2.1)

(4) Occasionally, we shall write $\underline{v}_e$ for the eulerian velocity, that is for the function $\underline{v}_e(\underline{x},t) = \underline{v}(\underline{X}(\underline{x},t),t)$.

it is assumed to hold for any subregion of C, we obtain the continuity equation:

a) in Lagrangean form

$$\rho_o = J\rho \quad ; \tag{2.2}$$

b) in mixed or Eulerian form

$$\frac{d\rho}{dt} + \rho \operatorname{div} \underline{U} = 0 \quad , \quad \frac{\partial\rho}{\partial t} + \operatorname{div}\rho\underline{U} = 0 \tag{2.3}$$

where $\frac{d\rho}{dt}$ is the molecular (or Lagrangean) derivative of ρ, while div is calculated with respect to the local variables $\underline{x}$.

Let $\underline{T}$ be the Cauchy stress tensor defined in the istantaneous configuration by

$$\underline{t}_n = \underline{T}\ \underline{n} \tag{2.4}$$

where $\underline{t}_n$ is the traction acting on the unit surface whose normal is $\underline{n}$ in the deformed configuration B_t; if $\rho\ \underline{b}$ are the body forces per unit volume in the deformed configuration, the principle of the conservation of the linear momentum is expressed by

$$\int_b \rho\ \underline{b}\ dv + \int_{\partial b} \underline{T}\ \underline{n}\ d\sigma - \frac{d}{dt}\int_b \rho\ \underline{v}\ dv = \underline{0}\ . \tag{2.5}$$

If the integrands are continuous and (2.5) is to be valid for all subregions b of the body,

$$\rho\ \underline{b} + \operatorname{div}\underline{T} - \rho\ \underline{a} = \underline{0} \quad ^{(1)}. \tag{2.6}$$

---

(1) However, the latter condition is no more universally accepted. Neglecting this hypothesis, a "non-local" theory of continuous bodies can be developed. See f.i. Edelen, & Laws[1], Edelen, Green & Laws[2], Green & Naghdi[3].

This equation is of a mixed form, because $\underline{a}$ is expressed as the molecular derivative of $\underline{v}$ whereas div $\underline{T}$ is calculated with respect to the local coordinates $\underline{x}$. We can give (2.6) a molecular form in the following manner. Coming back to (2.2) and (1.11), the equation (2.5) can be written as

$$\int_b \rho_o \underline{b}\, dV + \int_{\partial b_o} J\, \underline{T}\, (\underline{F}^T)^{-1} N\, d\Sigma - \int_{b_o} \rho_o\, \underline{a}\, dV = 0. \quad (2.7)$$

If we introduce the tensor

$$\underline{T}_R = J\, \underline{T}\, (\underline{F}^T)^{-1} \quad (2.8)$$

we can obtain in the place of (2.6)

$$\rho_o \underline{b} + \mathrm{div}\, \underline{T}_R - \rho_o \underline{a} = \underline{0}\ , \quad (2.9)$$

where now div is calculated with respect to the coordinates $\underline{X}$.

The Piola-Kirchhoff stress tensor $\underline{T}_R$ is a partially Eulerian non-symmetric tensor whose Cartesian components are given by

$$T_R^{\,h\,H} = J\, T^{h\,k}\, X^H_{\ ,k}\ . \quad (2.10)$$

Occasionally, it will be convenient to introduce a Lagrangean tensor $\underline{T}_o$ defined by

$$\underline{T}_o = \underline{F}^{-1}\underline{T}_R = J\, \underline{F}^{-1}\, \underline{T}\, (\underline{F}^T)^{-1} \quad (2.11)$$

whose Cartesian components are

$$T^{H\,K} = X^H_{\ ,h} T_R^{\,h\,K} = J\, X^H_{\ ,h} X^K_{\ ,k} T^{h\,k} = T^{KH}\ .$$

With the aid of $\underline{T}_o$, (2.9) can be written in the form

$$\rho_o \underline{b} + \mathrm{Div}\, (\underline{F}^{-1}\underline{T}_o) = \rho_o\, \underline{a}\ . \quad (2.12)$$

The equation of motion (2.6) must be supplemented by initial conditions:

$$\underline{x}\ (t_o,\underline{X}) = \underline{X} + \underline{u}_o(\underline{X}) \quad , \quad \dot{\underline{x}}\ (t_o,X)=\underline{v}_o(\underline{X}), \qquad (2.13)$$

and the boundary conditions

$$\underline{T}\ \underline{n} = \underline{f}\ (\underline{x}_\sigma,t) \qquad (2.14)$$

where $\underline{f}$ is a known function of the points of ∂B in the istantaneous configuration B_t.

The condition (2.14) is eulerian in form; it is convenient to write down his lagrangean form. A particular but noticeable case is the case in which $\underline{f}$ reduces to a pressure on ∂B, $f = - p\ \underline{n}$. Then (2.14) con be written as

$$\int_{\partial B} (\underline{T} + p\ \underline{1})\ \underline{n}\ d\sigma = \underline{0} \qquad (2.15)$$

from which, in the reference configuration

$$\int_{\partial B_o} (\underline{T}_R + p\ J\ (\underline{F}^T)^{-1})\ \underline{N}\ d\Sigma = \underline{0}\ , \qquad (2.16)$$

and therefore, in the hypothesis of the continuity of the integrand,

$$\underline{T}_R\underline{N} = -\ p\ J\ (\underline{F}^T)^{-1}\ \underline{N} \qquad (2.17)$$

where p (if not a constant) must be expressed as a function of $\underline{X}_\Sigma$, that is, of the unknown deformation.

More generally, let $d\sigma$ and $d\Sigma$ be corresponding elements of the boundaries of B and B_o, respectively; as

$$(d\sigma)^2 = J^2(d\Sigma)^2(\underline{F}^T)^{-1}\underline{N}\cdot(\underline{F}^T)^{-1}\underline{N} = J^2(d\Sigma)^2\underline{N}\cdot\underline{C}^{-1}\underline{N}\ ,$$

from (2.14) in integral form

$$\int_{\partial B} (\underline{T}\,\underline{n} - \underline{f})\, d\sigma = 0$$

we obtain

$$\int_{\partial B_o} (\underline{T}_R\,\underline{N} - f\,J\sqrt{\underline{N}\cdot\underline{C}^{-1}\underline{N}})\, d\Sigma = 0 \,, \qquad (2.18)$$

and therefore

$$\underline{T}_R\underline{N} = \underline{f}_R \quad , \quad \underline{f}_R = J\sqrt{\underline{N}\cdot\underline{C}^{-1}\underline{N}}\;\underline{f} \,. \qquad (2.19)$$

We remark that condition (2.19) is only formally analogous to the usual traction condition of linear elasticity. Indeed, vector $\underline{f}_R$ is unknown on ∂B_o, so that (2.19) is a non-linear condition for the displacement $\underline{u} = \underline{x} - \underline{X}$. The form of (2.19) seems to impose a procedure of successive approximations for solving non linear problems of elastodynamics.

If the boundary of the body is constrained to be supported by a rigid body the reactions of the support as well as the contact surface are unknown. The problem is non-linear even in the field of linear elasticity. This problem was introduced and deeply investigated by Signorini (Signorini[5], Fichera[6]) in the field of elastostatics; a formulation in the field of the elastodynamics is, to my knowledge, still lacking.

I shall finally remind that the balance of the moment of the momentum reduces to an identity if the stress tensor is symmetric, as I shall assume henceforth.

Indeed the symmetry of the stress tensor amounts to assuming that any torque applied to the body is generated by applied forces, i.e., that the equation of balance of the moment of momentum is

$$\int_b \underline{p} x \rho \underline{b}\, dv + \int_{\partial b} \underline{p} x \underline{T} n\, d\sigma = \frac{d}{dt}\int_b \rho\, \underline{p} x \underline{v}\, dv, \qquad (2.20)$$

where $\underline{p} = \underline{x} - X_o$ is the position vector of $\underline{x}$ with respect to the fixed point X_o. If there exists a distribution of body and surface couples, (2.20) must be replaced by

$$\int_b (\underline{p} x \rho \underline{b} + \underline{m})\, dv + \int_{\partial b} (\underline{p} x T n + \underline{m}_\sigma) d\sigma = \frac{d}{dt} \int_b \rho \underline{p} x \underline{v}\, dv \qquad (2.21)$$

In wiew of a usually admitted theorem analogous to the Cauchy's Lemma for the stress

$$\underline{m}_\sigma = \underline{M}\ \underline{n} \qquad (2.22)$$

so that, being

$$\int_{\partial b} \underline{p} x \underline{T} \underline{n}\ d\sigma = \int_b \underline{p} x\ \mathrm{div} \underline{T}\ dv - \int_b 2\underline{\Omega}\ dv$$

where $2\underline{\Omega} = \sum_k \underline{T}\ \underline{e}_k x \underline{e}_k = \hat{\underline{T}}$ = axial vector of $\underline{T}$, the equation

$$\rho\ \underline{m} + \mathrm{div}\ \underline{M} + 2\underline{\Omega} = \underline{0} \qquad (2.23)$$

is obtained as a second Cauchy's equation, supplemented by the boundary condition

$$\underline{M}\ \underline{n} = \underline{\Psi} \quad \text{on } \partial B\ . \qquad (2.24)$$

However this equation is not completetely satisfactory, because rotational momentum has been neglected. The problem is very complicated, and I shall refer the reader to the volume by Truesdell and Noll for further informations. See also f.e. Nowacki[7] and Stojanovic[8].

It is a easy to obtain from the above calculations the theorem of the kinetic energy. Multiplying equation (2.9) by $\underline{v}$ and integrating over B_o, we obtain

$$\int_{B_o} \underline{v}\cdot \mathrm{Div}\underline{T}_R \, dV = \int_{B_o} \mathrm{Div}(\underline{T}_R{}^T \underline{v}) \, dV - \int_{B_o} \mathrm{tr}(\underline{T}_R{}^T \mathrm{Grad}\, \underline{v}) \, dV =$$

$$\int_{\partial B_o} \underline{v}\cdot \underline{f}_R \, d\Sigma - \int_{B_o} \mathrm{tr}(\underline{T}_R{}^T \dot{\underline{F}}) \, dV \quad ,$$

and therefore

$$\int_{B_o} \rho_o \underline{b}\cdot\underline{v} \, dV + \int_{\partial B_o} \underline{v}\cdot \underline{f}_R \, d\Sigma - \int_{B_o} \mathrm{tr}(\underline{T}_R{}^T \dot{\underline{F}}) \, dV = \frac{d}{dt}\frac{1}{2}\int_{B_o} \rho_o \underline{v}^2 dV$$

or

$$\frac{dT}{dt} = P + W \ , \tag{2.25}$$

where

$$T = \frac{1}{2}\int_{B_o} \rho_o v^2 dV, \quad W = \int_{B_o} \rho_o \underline{b}\cdot\underline{v} dV + \int_{\partial B_o} \underline{v}\cdot \underline{f}_R d\Sigma, P = -\int_{B_o} \mathrm{tr}(\underline{T}_R^T \dot{\underline{F}}) dV \quad (2) \tag{2.26}$$

(2) Of course, we can also obtain $W = \int_B \rho \underline{b}\cdot\underline{v} \, dv + \int_{\partial B} \underline{f}\cdot\underline{v} \, d$ because of $\underline{f}_R = \frac{d\sigma}{d\Sigma}\underline{f}$.

3 - SMOOTH DEFORMATION; EVERSIONS.

In n. 1 a smooth deformation was defined as an one-to-one correspondence between B_o and B for which J is strictly positive; a different, more restrictive definition is possible, perhaps endowed with a more clear physical meaning.

Let us suppose that the deformation is a continuous function of a real parameter $\lambda \varepsilon [0,1]$ with B_o corresponding to the value $\lambda = 0$ of the parameter, and B to the value $\lambda = 1$. We can therefore consider $\underline{x}$ as a function (of $\underline{X}$ and) of λ, writing

$$\underline{x} = \underline{\chi}_\lambda (\underline{X}) \tag{3.1}$$

with

$$\underline{\chi}_o(\underline{X}) = \underline{X} \quad , \quad \underline{\chi}_1(\underline{X}) = \underline{x} \tag{3.2}$$

and

$$\text{Grad} \ \underline{\chi}_\lambda(\underline{X}) = \underline{F}_\lambda . \tag{3.3}$$

Throughout this Section each deformation for which $J(\lambda) > 0$ for any λ will be called regular.

It is clear that this definition is more restrictive that the one adopted previously; this is evident from the following example.

Let C be an hollow cylinder, with inner radius R_i and outer radius R_e, $R_i < R_e$, and let us consider the finite deformation defined, in cylindrical coordinates, by

$$r = \sqrt{A R^2 + B} \qquad \theta = \Theta \ , \quad z = F Z \tag{3.4}$$

with A, B and F constants restricted by the condition $r > 0$.

It is easy to obtain

$$J = A F \tag{3.5}$$

and therefore, if A F > 0, the deformation is regular in the sense of n. 1.

The inner and the auter radii in the deformed configuration are

$$r_i = \sqrt{A\,R_i^2 + B} \quad , \quad r_e = \sqrt{A\,R_e^2 + B} \,. \tag{3.6}$$

If A > 0, r_i is smaller that r_e, but it is also possible to make $r_i < r_e$ for other choices of A. Indeed, to this end it is sufficient that

$$A\,R_i^2 + B \;>\; A\,R_e^2 + B \quad ,$$

which implies a < 0. Therefore, if F < 0, the deformation is smooth and the cylinder is subjected to an eversion.

However, let us now consider a continuous deformation from B_o to B continuously depending on a parameter λ. A, B, and F are therefore continuous functions of λ and the conditions $r_{\lambda=0} = R$, $z_{\lambda=0} = Z$ implies A(0) = 1, B(0) = 0, F(0) = 0. Now, if $r_i > r_e$, A must be negative for λ = 1, and therefore there exists a value of λ, say $\bar{\lambda}$, such that $A(\bar{\lambda}) = 0$ and therefore $J(\bar{\lambda}) = 0$: the deformation is no more regular in the sense previously explained.

A more sophisticated example has been occasionally elaborated by P. Podio Guidugli in a paper still now unpublished. (Podio Guidugli [9]).

I can give now an example of a non smooth motion.

Let us consider an acceleration-less motion of a continuous body. Therefore,

$$\underline{x} = \underline{\phi}(\underline{X})\; t + \underline{X} \tag{3.7}$$

where $\underline{\phi}(\underline{X}) = \dot{\underline{x}}$ = const. in time. From (3.7) and $\underline{x} = \underline{\chi}(\underline{X},t)$,

$$\underline{x} = \underline{X}(\underline{x} - \phi\,(\underline{X})\,t\,,\,t) \tag{3.8}$$

a functional relation for $\underline{x}$ due to B. Caldonazzo who, first, studied this kind of motion (Caldonazzo[10]).

I shall consider in particular a motion for which

$$\dot{\underline{x}} = \underline{a}(t)\,\underline{x} + \underline{b}(t)\ . \tag{3.9}$$

If the motion must be accelerationless, $\underline{a}$ and $\underline{b}$ must satisfy the equations

$$\dot{\underline{a}} + \underline{a}^2 = \underline{0}\quad,\quad \dot{\underline{b}} + \underline{a}\,\underline{b} = \underline{0}\ . \tag{3.10}$$

The general solution of (3.10) is given by

$$(\underline{1} + \underline{A}\,t)\,\underline{a} = \underline{A}\ ,\ (\underline{1} + \underline{A}\,t)\,\underline{b} = \underline{B} \tag{3.11}$$

where $\underline{A} = \underline{a}\,(t_o)$ and $\underline{B} = \underline{b}\,(t_o)$; (3.11) shows that the motion can be smooth only in a finite interval of time [1].

A dynamical problem similar to the kinematical problem just examinated is the following.

Let $\underline{b}$ be a constant vector, and let the body be homogeneous in the reference configuration, and $\underline{T}$ be a function of F and

---

(1) Let us consider the identy $\underline{a}^{-1}\underline{a} = \underline{1}$. The derivative of this identity gives

$$\dot{\underline{a}}^{-1}\,\underline{a} + \underline{a}^{-1}\,\dot{\underline{a}} = 0$$

from which

$$\dot{\underline{a}}^{-1} = -\,\underline{a}^{-1}\ \underline{a}\,\underline{a}^{-1}\ .$$

But, from (3.10) $\dot{\underline{a}} = -\,a^2$ and therefore $\dot{\underline{a}}^{-1} = \underline{1}$, from which (3.11).

$\underline{X}$. We shall look for a solution of the dynamical equation

$$\rho \underline{b} + \operatorname{div} \underline{T} = \rho \underline{a}$$

corresponding to a homogeneous motion, $F = F(t)$, for which therefore

$$\underline{x}(t,X) = \underline{x}_o(t) + F(t)\ (\underline{X} - \underline{X}_o)\ . \qquad (3.12)$$

If a motion of this kind is possible, B is homogeneous also in the istantaneous configuration B, and T is independent of $\underline{x}$. Therefore, the equation of motion gives

$$\ddot{\underline{x}}_o + \ddot{\underline{F}}\ (\underline{X} - \underline{X}_o) = \underline{b}$$

which implies, as $\underline{X}$ is arbitrary,

$$\ddot{\underline{x}}_o = \underline{b}\quad ,\quad \ddot{\underline{F}} = \underline{0}$$

that is

$$\underline{x}_o = \frac{1}{2}\,\underline{b}\,t^2 + \underline{v}_o t + \underline{X}_o\quad ,\quad \underline{F} = \underline{F}_o(1 + t\,\underline{F}_1)\ . \qquad (3.13)$$

Now, det $\underline{F}$ must be greater than zero for all t which amounts to the condition

$$\det\,(\underline{1} + t\,\underline{F}_1) > 0$$

If $\underline{F}_1$ possesses a negative proper number, this condition is violated for some t, and therefore the motion is not smooth for all t.

A different kind of non-regular motion is the motion classified as wave propagation.

Let the body be indefinitely extended, and let S_t be a smooth surface in motion with respect to the istantaneous configuration B of the body; S_t is called a wave if some of the functions which characterize the motion of B is discontinuous across S_t. More precisely, let ψ be any such discontinuous function; if ψ^+ (ψ^-) is the limit of ψ when $\underline{x}$ tends to S_t along the positive (negative) normal, $\psi^+ \neq \psi^-$ and $\psi^+ - \psi^- = [\psi]$ is the discontinuity of ψ across S_t. Of course, $[\psi]$ is a function of $\underline{x}$ whose domain is S_t itself. In continuum mechanics, if ψ is the acceleration of the body, the wave is called an acceleration wave, while, if ψ is the velocity the wave is a shock (2).

It is clear that discontinuities must satisfy some conditions of compatibility due to the nature of discontinuities themselves. These conditions are called the Hugoniot-Hadamard conditions, or the kinematical conditions, and must be satisfied independently of the material. A different kind of conditions emerges when dynamical equations are considered, which must be identically satisfied in the two subspaces in which S_t cuts out B. These dynamical conditions together with the Hugoniot-Hadamard conditions,

are only necessary conditions for the propagation to be possible, and in general are very complicated in form. The problem appears particularly complicated for shock waves, and has been extensively investigated only for unidimensional waves.

---

(2) However, the discontinuity of the velocity must be non--tangent to S_t.

REFERENCES

1. Edelen, D.G.B. & Laws, N., On the thermodynamics of systems with non locality, *Arch. Rat. Mech. An.*, 43, 24, 1971.
2. Edelen, D.G.B., Green, A.E. & Laws, N., Non local continuum mechanics, *Arch. Rat. Mech. An.*, 43, 36, 1971.
3. Green, A.E. & Naghdi, P.M., On continuum thermodynamics, *Arch. Rat. Mech. An.*, 48, 352, 1972.
4. Signorini, A., Trasformazioni termoelastiche finite, Mem. IV, *Ann. Mat. pura appl.*, (4), 51, 329, 1960.
5. Signorini, A., Questioni di elasticità non linearizzata e semilinearizzata, *Rend. di Mat.*, 18, 95, 1959.
6. Fichera, G., Boundary value problems with unilateral constraints, in *Encyclopedia of Physics*, Vol. VI a/2, Springer, Berlin, 1972.
7. Nowacki, W., *Theory of micropolar elasticity*, CISM, Udine, 1970.
8. Stojanovic, R., *Recent developments in the theory of polar continua*, CISM, Udine, 1970.
9. Podio Guidugli, P., De Giorgi's counterexample in elasticity, *Quart. Appl. Math.* (in press), 1975.
10. Caldonazzo, B., Sui moti liberi di un mezzo continuo, *Ann. Mat. pura appl.* (4), 26, 43, 1947.
11. Chen, P.J., Growth and decay of waves in solids, in *Encyclopedia of Physics*, Vol. VI a/3, Springer, Berlin, 1973.

CHAPTER II

THE PRINCIPLES OF CLASSICAL THERMODYNAMICS. THERMOELASTIC BODIES

1 - THE PRINCIPLES OF CLASSICAL TERMODYNAMICS.

The balance equations

$$\frac{d\rho}{dt} + \rho \operatorname{div}\underline{v} = 0 \quad , \rho\, \underline{b} + \operatorname{div} \underline{T} - \rho\, \underline{a} = 0 \tag{1.1}$$

are an undetermined system of partial differential equations for the unknowns ρ and $\underline{u}$. These equations are to be supplemented with constitutive equations. Usually is produced some phenomonological constitutive equation whose form is suggested by physical experiments. However, as the presence of some thermodinamical variable is often unavoidable, it seems preferable to begin with the classical principles of the thermodynamics.

Let Θ be the absolute temperature of the particles of the body; the first principle of thermodynamics in its classical formulation, consists in assuming the existence of a function E, the internal energy of the body, and of a thermical power Q such that the balance equation

$$\frac{dE}{dt} + \frac{dT}{dt} = W + Q \tag{1.2}$$

is satisfied along each possible thermodynamical process. In (1.2), T is the kinetic energy of the body, and W the mechanical power of the external forces acting on the body. Of course, T is connected with W by the balance equation for mechanical energy [see (2.20, 21) of the Ch. I]

$$W = - \int_{B_o} tr(\underline{T}_R^T \, \dot{\underline{F}}) \, dV = \frac{dT}{dt} \, , \tag{1.3}$$

so that equation (1.2) becomes

$$\frac{dE}{dt} = Q + \int_{B_o} tr(\underline{T}_R^T \, \dot{\underline{F}}) \, dV = Q - P \, , \tag{1.4}$$

here P is the stress power, $P = - \int_{B_o} tr(\underline{T}_R^T \, \dot{\underline{F}}) \, dV$ (1).

Now, if E has the density ε

$$E = \int_B \rho\varepsilon dv = \int_{B_o} \rho_o \varepsilon \, dV \tag{1.5}$$

and the heat power is the sum of a heat flux across the boundary of the body and a local heat supply, so that

$$Q = \int_{\partial B} \underline{q} \cdot \underline{n} \, d\sigma + \int_B \rho s \, dv = \int_{\partial B_o} \underline{q}_R \cdot \underline{N} \, d\Sigma + \int_{B_o} \rho_o \, s \, dV \tag{1.6}$$

where

$$\underline{q}_R = J \, \underline{F}^{-1} \, \underline{q} \tag{1.7}$$

we can obtain, under the usual hypothesis of continuity and the validity of (1.4) for all partitions of B_o, the local equation of balance

$$\rho_o \frac{d\varepsilon}{dt} = \mathrm{Div} \, \underline{q}_R + \rho_o \, s - w_R \tag{1.8}$$

(1) P can also be expressed by $P = - \int_B tr(\underline{T} \, \underline{D}) \, dv$ where

$$D_{hk} = \frac{1}{2} (v_{h,k} + v_{k,h})$$

where

$$w_R = -\operatorname{tr}(\underline{T}_R^T\,\dot{\underline{F}}) \quad [2] \tag{1.9}$$

The formulation of the second principle is much more difficult. In simpler form it is equivalent to the existence of another function of state for the body, the entropy H, whose internal production is non-negative for all admissible processes [3].

If the external production of entropy is generated by a distribution of sources in the body and by an entropy flux across the boundary, we can express the second principle in the form

$$\frac{dH}{dt} - \int_B \rho r\, dV - \int_{\partial B} h\, d\sigma \geq 0. \tag{1.10}$$

It is customary to assume for r and h the form

$$r = s/\Theta \quad , \qquad h = \underline{q}\cdot n/\Theta$$

If η denotes the density of entropy, then equation (1.10) can be put in the form

$$\frac{d}{dt}\int_B \rho\eta dv - \int_B \left(\operatorname{div}\frac{q}{\Theta} + \rho\,\frac{s}{\Theta}\right) dv \geq 0\ , \tag{1.11}$$

and under the usual assumptions of continuity, in the local form

---

(2) W_R is also given by $w_R = -\frac{1}{2}\operatorname{tr}(\underline{T}_o\,\dot{\underline{C}})$.

(3) A thermodynamic process, in the present context, is a pair of functions of t, $\theta = \theta(X,t)$, $\underline{x} = \underline{x}(X,t)$ sufficiently smooths, $\theta > 0$.

$$\rho\dot{\eta} - \rho\,\frac{s}{\Theta} - \frac{1}{\Theta}\,\mathrm{div}\,\underline{q} + \frac{1}{\Theta^2}\,\underline{q}\cdot\mathrm{grad}\,\Theta \geq 0 \tag{1.12}$$

which must be satisfied in all processes. Inequality (1.12) can be easily given a molecular form

$$\rho_o\dot{\eta} - \rho_o\,\frac{s}{\Theta} - \frac{1}{\Theta}\,\mathrm{div}\,\underline{q}_R + \frac{1}{\Theta^2}\;\underline{q}\cdot\mathrm{grad}\,\Theta \geq 0 \tag{1.13}$$

Let us introduce the Gibbs free energy ψ, defined by

$$\psi = \varepsilon - \eta\Theta \quad ; \tag{1.14}$$

from (1.13) and (1.8) we obtain

$$\rho_o\dot{\psi} + \rho_o\eta\dot{\Theta} + w_r - \frac{1}{\Theta}\,\underline{q}_R\cdot\mathrm{grad}\,\Theta \leq 0 \tag{1.15}$$

where the dot stands for molecular time derivative.

The assumption that (1.15) is satisfied in all possible processes results in severe restrictions for the constitutive equations. Laying aside the general considerations, I shall now define a thermoelastic body as a body for which: a) a reference configuration B_τ exists at an uniform temperature τ such that for each process starting from B_τ the internal entropy production is zero

$$\rho_o\dot{\psi} + \rho_o\eta\dot{\Theta} + w_r = 0 \tag{1.16}$$

b) the free energy, the specific entropy and the stress tensor $\underline{T}$ are functions only of the temperature and the deformation gradient.

For a thermoelastic solid we obtain at once from (1.15)

$$\underline{q}_R\cdot\mathrm{grad}\,\Theta \geq 0 \tag{1.17}$$

or, equivalently,

$$\underline{q} \cdot \text{grad } \Theta \geq 0. \tag{1.18}$$

On the other hand, from (1.16) we obtain as constitutive equations for a thermoelastic body

$$\eta = -\frac{\partial\psi}{\partial\Theta} \quad , \quad T_R{}^H{}_h = \rho_o \frac{\partial\psi}{\partial x^h{}_{,H}} \tag{1.19}$$

or, for the Lagrangean stress tensor

$$\underline{T}_o = \rho_o \partial_{\underline{E}}\psi \ , \quad T^{HK} = \frac{1}{2}\rho_o\left(\frac{\partial\psi}{\partial E_{HK}} + \frac{\partial\psi}{\partial E_{KH}}\right) . \tag{1.19'}$$

The usual assumptions concerning the specific heat at constant configuration assure the one-to-one corrispondence between η and Θ, as the second derivative of ψ with respect to the temperature must be different from zero. We can also remark that the knowledge of the function ψ assures complete knowledge of the specific entropy and of the Lagrangean stress.

Particular, but important, processes are the isothermal and the isentropic processes. For the former, $\Theta = \tau$, so that, if we put

$$W_\tau = \rho_o\{\psi(\underline{E},\tau,\tau;\underline{X}) - \psi(0,\tau,\tau;\underline{X})\} \tag{1.20}$$

(1.19') can be put in the form

$$\underline{T}_o = \partial_E W_\tau \tag{1.21}$$

The function W_τ is the isothermal potential of the stress(4)

(4) It must be remarked that the Lagrangean stress tensor is a function (of τ and) of the deformation tensor $\underline{E}$, while the Cauchy stress tensor $\underline{T}$ is a function of the local rotation $\underline{R}$ as well.

On the other hand, if the process is isentropic, it is convenient to write (1.16), in the form

$$\rho_o \Theta\dot{\eta} - \rho_o \varepsilon - W_R = 0 . \quad (1.22)$$

Putting

$$W_\eta = \rho_o \{\varepsilon(\underline{E},\eta,\eta;\underline{X}) - \varepsilon(0,\eta,\eta;\underline{X})\} \quad (1.23)$$

we therefore obtain

$$\Theta = \frac{\partial\varepsilon}{\partial\eta} \quad , \quad T^{HK} = \frac{1}{2}\rho_o\left(\frac{\partial\varepsilon}{\partial E_{KH}} + \frac{\partial\varepsilon}{\partial E_{KH}}\right) = \frac{1}{2}\left(\frac{\partial W_\eta}{\partial E_{HK}} + \frac{\partial W_\eta}{\partial E_{KH}}\right) \quad (1.24)$$

The function W_η is the isentropic potential of the stress.

2 - MÜLLER'S GENERALIZATION OF THE SECOND PRINCIPLE.

The formulation of the second principle is essentially based upon two different axioms: a) the internal production of entropy is never negative; b) the external sources of entropy have densities s/Θ and $\underline{q}\cdot n/\Theta$.

Müller proposed to retain the first axiom, and to substitute the second with the assumption

$$\rho_o\dot{\eta} + \text{div}\,\Phi \geq 0 \quad , \quad (2.1)$$

where Φ is a vector of constitutive nature. In his papers on this subjected, Müller deeply investigated the consequences of the inequality (2.1). He assumed η and Φ to be functions of $\underline{F}$, θ, $\dot{\theta}$ and grad θ, where θ is the empirical temperature of the points of the body, and among other things he was able to define the absolute temperature and to explain some features of the heat equation which are not satisfactory from a physical point of view (Muller[1,2,3]. See also Green & Laws[4], Green & Lindsay[5], Gurtin & Williams[6], Coleman & Owen[7], Warren & Chen[8]).

3 - INTERNAL CONSTRAINTS.

We shall consider only constraints of the form

$$\Phi\ (\underline{E},\Theta) = 0 \quad (^1) \tag{3.1}$$

or, equivalently

$$\mathrm{tr}\ (\partial_{\underline{E}}\Phi\ \dot{\underline{E}}) + \frac{\partial\Phi}{\partial\Theta}\ \dot{\Theta} = 0 \quad . \tag{3.2}$$

For a thermoelastic solid, (1.16) and (3.2) must be identically satisfied. Let $-p$ be the Lagrangean multiplier; one obtains

$$\rho_o\dot{\psi} + (\rho_o\eta - p\ \frac{\partial\Phi}{\partial\Theta})\dot{\Theta} - p\ \mathrm{trac}(\partial_{\underline{E}}\Phi\ \dot{\underline{E}}) + W_R = 0 \quad , \tag{3.3}$$

from which follows the constitutive equations

$$\eta = -\ p\ \frac{\partial\Phi}{\partial\Theta} - \frac{\partial\psi}{\partial\Theta} \quad , \quad \underline{T}_o = -\ p\ \partial_{\underline{E}}\Phi + \rho_o\partial_{\underline{E}}\psi \quad . \tag{3.4}$$

(^{1}) A constraint of the form $\Phi(\underline{F},\Theta) = 0$ cannot be admissible. In fact, Φ is a function whose values are scalars; therefore, Φ must be invariant under rigid rotations, that is

$$\Phi\ (\underline{Q}\ \underline{F},\ \Theta) = \Phi\ (\underline{F},\ \Theta)$$

for all orthogonal $\underline{Q}$. Choosing $\underline{Q} = \underline{R}^T$, (3.1) follows.

The multiplier p characterizes the internal reaction, while the constitutive equation $\psi = \psi(\underline{E}, \Theta)$ no longer completely specifies η and $\underline{T}_o$.

Equation (3.4) implies

$$\underline{T} = \frac{1}{J}\,\underline{F}\,\underline{T}_o\underline{F}^T = -\frac{p}{J}\,\underline{F}\,\partial_{\underline{E}}\Phi\,\underline{F}^T + \rho_o\,\underline{F}\partial_{\underline{E}}\psi\underline{F}^T \quad . \tag{3.5}$$

If we chose

$$\Phi = J - 1 \tag{3.6}$$

we obtain the particular but noticeable case, when

$$\underline{T} = -\,p\,\underline{1} + \rho_o\,\underline{F}\partial_{\underline{E}}\psi\,\underline{F}^T \tag{3.7}$$

For a Signorini's thermoelastic incompressible solid (Signorini[9])

$$J - f(\Theta,\tau) = 0 \ , \tag{3.8}$$

with $f(\tau,\tau) = 1$, in place of (3.4) we obtain

$$\eta = -\frac{p}{\rho_o}\frac{\partial f}{\partial \Theta} - \frac{\partial \psi}{\partial \Theta} \ , \quad \underline{T} = -\,p\,\underline{1} + \rho_o\underline{F}\,\partial_{\underline{E}}\psi\underline{F}^T \tag{3.9}$$ (2)

(2) In a theory more general than that exposed here, the constitutive equations can be expressed by means of functionals of the form

$$T = \mathcal{F}\,(\underline{X},\underline{F}^t(s)\,)$$

$\underline{F}^t(\underline{X},s)=\underline{F}(\underline{X},t-s), 0\leq s<+\infty$ is the history of deformation to the time t. In such a theory, it seems reasonable to introduce constraints expressed by means of functional relations of the form $\Gamma(\underline{X},\underline{F}^t\,(s)\,) = 0$. However, under the hypotheses usually accepted in the theory of simple bodies, it can be proved that Γ must reduce to an ordinary function of the instantaneous determination of $\underline{F}$. A general theory on this subject was developed by M.E. Gurtin and P. Podio Guidugli. (Gurtin & Podio Guidugli[9]. See also Manacorda[10]).

4 - ISOTROPIC HOMOGENEOUS BODIES.

I shall lay aside the general theories of the symmetry group for a material, and I shall give here a simplified definition of an isotropic body.

By definition a thermoelastic body is isotropic (in some reference configuration B_τ) if the free energy of the body is a function (of τ, the instantaneous temperature Θ and) of the three principal invariants of $\underline{E}$, i.e.

$$\psi\ (\underline{E},\ \Theta,\tau;\underline{X}) = \psi\ (I_{\underline{E}},\ II_{\underline{E}},\ III_{\underline{E}},\ \Theta,\ \tau;\ X). \qquad (4.1)$$

An isotropic body is homogeneous in C_τ if ρ_o and ψ do not depend explicitely on X.

Discuss now how can a reference configurations preserving isotropy and homogeneity be characterized. Let $\underline{P}'$ be the gradient of the deformation from B_o to B_o'. Because $\rho_o=\rho_o'$ det $\underline{P}'$, therefore ρ_o and ρ_o' do not depend of $\underline{X}$ and det $\underline{P}'$ must likewise be independent of $\underline{X}$. Moreover, the free energy of the body, as a function of the deformation tensor $\underline{E}'$, must be an isotropic function. For this to be the case, a necessary and sufficient condition is that the deformation $B_o \to B_o'$ reduce to a conformal transformation, that is, to an uniform dilatation whose parameter is independent of $\underline{X}$. Hence

$$\underline{P}' = \lambda\ \underline{1}\ ,$$

with λ independent of $\underline{X}$ (Signorini[11]).

Henceforth, the reference configuration in which the material is homogeneous and isotropic will be called natural. It can be proved that a natural equilibrium configuration is free of stresses.

I shall now remind the various measures of deformation already introduced:

$$\underline{F}\,\underline{F}^T = \underline{C} = \underline{1} + 2\,\underline{E} = \underline{U}^2 \qquad ,\underline{F} = \underline{R}\,\underline{U}$$

$$\underline{F}^T\,\underline{F} = \underline{B} = \underline{c}^{-1} = (\underline{1} + 2\underline{e})^{-1} = \underline{V}^2, \underline{F} = \underline{V}\,\underline{R} \tag{4.2}$$

$$\underline{B} = \underline{R}\,\underline{C}\,\underline{R}^T .$$

The invariants of $\underline{B}$ and $\underline{C}$ coincide and are in one-to-one correspondence with the invariants of $\underline{c}$ (or of $\underline{1} + 2\underline{e}$). In fact

$$I_{\underline{c}} = \frac{II_{\underline{c}^{-1}}}{III_{\underline{c}}} = \frac{II_{\underline{B}}}{J^2} = \frac{II_{\underline{C}}}{J^2} , \quad II_{\underline{c}} = \frac{I_{\underline{c}^{-1}}}{III_{\underline{c}^{-1}}} = \frac{I_{\underline{B}}}{J^2} = \frac{I_{\underline{C}}}{J^2} ,$$

$$III_{\underline{c}} = \frac{1}{J^2} \tag{4.3}$$

Therefore, we shall consider ψ as a function of the invari̲ants of any one of the tensors just listed. We can thus obtain various forms of the stress-strain law for an isotropic material;

$$\underset{o}{\underline{T}} = L_{\underline{E}}\,\underline{1} + M_{\underline{E}}\,\underline{E} + N_{\underline{E}}\,\underline{E}^2 = L_{\underline{C}}\,1 + M_{\underline{C}}\underline{C} + N_{\underline{C}}\underline{C}^2 , \tag{4.4}$$

where L, M, N etc. are functions (of the temperature and) of the invariants of $\underline{E}$ or of $\underline{C}$.

Now, as $\underline{T} = J^{-1}\,\underline{F}\,\underline{T}\,\underline{F}^T$, we obtain for the Cauchy stress

$$\underline{T} = \ell_{\underline{C}}\underline{1} + m_{\underline{C}}\underline{C}_R + n_{\underline{C}}\underline{C}_R^2 = \ell_{\underline{B}}\underline{1} + m_{\underline{B}}\underline{B} + n_{\underline{B}}\underline{B}^2 , \tag{4.5}$$

with

$$\underline{C}_R = \underline{R}\,\underline{C}\,\underline{R}^{-1} = \underline{B} \tag{4.6}$$

Using the Cayley-Hamilton identity[2] the last relation may be written in the form

$$\underline{T} = \alpha\, \underline{1} + \beta\, \underline{B} + \gamma\, B^{-1} \tag{4.7}$$

or in the form

$$\underline{T} = \ell_e \underline{1} + m_e \underline{e} + n_e \underline{e}^2 \ . \tag{4.8}$$

Let the body be isotropic and let B_o be a natural reference configuration of equilibrium in which the external forces are zero, $\underline{b} = \underline{0}$, $\underline{f} = \underline{0}$; the equilibrium equations reduce to

$$\operatorname{div} \underline{T} = \underline{0} \text{ in } B_o \ , \ \underline{T}\, \underline{N} = \underline{0} \quad \text{on } \partial B_o \ . \tag{4.9}$$

Now, in B_o, $\underline{E} = \underline{0}$, or $\underline{C} = \underline{B} = \underline{1}$, so that $\underline{T}$ reduces to an isotropic tensor $\underline{T} = -\, p\, \underline{1}$, where p must satisfy the equation

$$\operatorname{div}(p\, \underline{1}) = \operatorname{grad} p = \underline{0} \text{ in } B_o.$$

Therefore, p = cost. in B_o; but, on the boundary, $p\, \underline{N} = \underline{0}$, which implies p = 0: the natural configuration B_o for the body is therefore free of stresses.

(2) The Cayley-Hamilton identity for $\underline{B}$ is: $-\underline{B}^3 + I_B \underline{B}^2 - II_B \underline{B} +$ $+ III_B \underline{1} = \underline{0}$, or rather $-\underline{B}^2 + I_B \underline{B} - II_B \underline{1} + III_B \underline{B}^{-1} = \underline{0}$ delivering $\underline{B}^2$ as a function of $\underline{B}$ and of $\underline{B}^{-1}$.

REFERENCES

1. Müller, I. *Entropy, absolute temperature and coldness in Thermodynamics*, CISM, Udine, 1971.

2. Müller, I., *Thermodynamik. Die Grundlagen der Materialtheorie*, Bartelsmann, Dortmund, 1973.

3. Müller, I., Die Kältefunktion, eine universelle Funktion in die Thermodynamik viskoser Wärmeleitender, *Arch. Rat. Mech. An.*, 40, 1, 1971.

4. Green, A.E. & Laws, N., On a global entropy production inequality, *Q.J.Mech. appl. Math.*, 25, 1, 1972.

5. Green, A.E. & Lindsay, K.A., Thermoelasticity, *J. of Elast.*, 2, 1, 1972.

6. Gurtin, M.E. & Williams, W.O., An axiomatic foundation for continuum thermodynamics, *Arch. Rat. Mech. An.*,26,83,1967.

7. Coleman, B.D. & Owen D.R., A mathematical foundation for Thermodynamics, *Arch. Rat. Mech. An.*, 54, 1, 1974.

8. Warren, E.W. & Chen, P.J., Wave propagation in the two-temperature theory of thermoelasticity, *Acta Mech.* 16,21,1973.

9. Gurtin, M.E. & Podio Guidugli, P., The thermodynamics of constrained materials, *Arch. Rat. Mech. An.*, 51, 3, 1973.

10. Manacorda, T., Una osservazione al riguardo dei vincoli interni in un solido, *Riv. Univ. Parma* (in press).

11. Signorini, A., Trasformazioni termoelastiche finite, Mem. III, *Ann. Mat. pura appl.* (4), 39, 147, 1955.

CHAPTER III

THÉ ELASTIC POTENTIAL. TOLOTTI'S THEOREM

1 - TOLOTTI'S THEOREM.

The free energy of a thermoelastic solid is a function of the reference configuration and of the reference temperature τ. It seems reasonable that the form of the function ψ must be independent of τ when B_τ varies in the set of the natural reference configurations of a given isotropic homogeneous body. More precisely, assume that for each reference temperature $\theta \epsilon [\tau_1, \tau_2]$ a natural reference configuration exists and, following Tolotti (Tolotti [1]) assume that the function ψ is independent of τ, namely

$$\psi\,(\underline{E}', \Theta, \tau') = \psi\,(\underline{E}, \Theta, \tau) \quad ^{(1)} \tag{1.1}$$

where $\underline{E}'$ is the deformation tensor for the deformation $B_{\tau'} \to B$. However it will be convenient to express ψ as a function of C instead of $\underline{E}$.

For the sake of simplicity, put

$$Y_1 = I_C\;,\; Y_2 = II_C\;,\; Y_3 = J = \sqrt{III_C} \tag{1.2}$$

so that Tolotti's hypothesis can be written in the form

$$\psi\,(\underline{Y} \,|\, \theta\,,\, \tau') = \psi\,(\underline{Y} \,|\, \theta\,,\, \tau) \tag{1.3}$$

---

(1) θ can be different for different points $\underline{x}$. We shall consider a fixed point $\underline{x}$.

or, equivalently

$$\frac{\partial\psi}{\partial\tau} + \Sigma_i \frac{\partial\psi}{\partial Y_i} \frac{\partial Y_i}{\partial\tau} = 0 \qquad (1.4)$$

a partial differential equation for ψ. Viceversa, each solution of (1.4) satisfies Tolotti's hypothesis (2).

Let us put now

$$\ell_\tau = \rho_\tau^{-1/3} \qquad (1.5)$$

(where $\rho_\tau = \rho_o$ is the density in the reference configuration) so that the parameter of the uniform dilatation connecting B_τ with $B'_{\tau'}$ is given by

$$d = \frac{\ell_{\tau'}}{\ell_\tau} = \left(\frac{\rho_\tau}{\rho_{\tau'}}\right)^{1/3} \qquad (1.6)$$

Taking into account the relation between $\underline{C}'$ and $\underline{C}$

$$\underline{C}' = \underline{P}^T\, \underline{C}\, \underline{P} = d^{-2}\, \underline{Q}\, \underline{C}\, \underline{Q}^T \quad ,$$

where $\underline{Q}$ is a rigid proper rotation, we obtain for the principal invariants of $\underline{C}$ and $\underline{C}'$ (3).

$$\ell_{\tau'}^2\, Y_1' = \ell_\tau^2\, Y_1, \quad \ell_{\tau'}^4\, Y_2' = \ell_\tau^4\, Y_2, \quad \ell_{\tau'}^3\, Y_3' = \ell_\tau^3\, Y_3 \qquad (1.7)$$

which evidently imply that the three variables

$$Z_1 = \ell_\tau^2\, Y_1 \ , \quad Z_2 = \ell_\tau^4\, Y_2 \ , \quad Z_3 = \ell_\tau^3\, Y_3$$

---

(2) ψ is defined within a constant.

(3) That is, for the invariants relative to the deformations $B_\tau \to B$ and $B_{\tau'} \to B$, respectively.

are invariant under the deformation $B_\tau \to B_{\tau'}$.

Now, the most general solution of (1.4) is a function of three independent solutions. Therefore, the most general ψ is a function of Z_1, Z_2, Z_3 namely

$$\psi\ (Y \mid \Theta,\tau) = \hat{\psi}(Z \mid \Theta)\ . \tag{1.8}$$

In particular, ψ may depend on τ only via ℓ_τ, that is of ρ_τ.

Putting $\tau' = \Theta$ in (1.7) and solving for Y_I, Y_2, Y_3, we obtain

$$Y_1^\Theta = \frac{\ell_\tau^2}{\ell^2}\, Y_1\ ,\quad Y_2^\Theta = \frac{\ell_\tau^4}{\ell^4}\, Y_2\ ,\quad Y_3^\Theta = \frac{\ell_\tau^3}{\ell^3}\, Y_3 \tag{1.9}$$

where $Y^\Theta = Y'$ for $\tau' = \Theta$ and $\ell = \ell_{\tau'}$ for $\tau' = \Theta$. Now, for any Θ, the Y's are proportional to the Z's; we can infer that ψ must be a function of Θ and Y^Θ

$$\psi\ (Y|\Theta,\tau) = \bar{\psi}\ (Y^\Theta|\Theta) \quad (^4) \tag{1.10}$$

We can also come back to the definition of the isothermal potential [Ch. II (1.20)] at the temperature τ

$$W_\tau = \rho_\tau\ \{\psi\,(Y^\tau|\tau,\tau) - \psi\,(Y_o|\tau,\tau)\} \tag{1.11}$$

where $Y_o = (0,0,1)$. Solving for ψ

$$\psi\ (Y^\tau|\tau,\tau) = \bar{\psi}\ (Y^\tau|\tau) = \frac{1}{\rho_\tau}\, W_\tau + \psi\ (Y_o|\tau,\tau)$$

and putting $\tau = 0$

(^{4}) $\bar{\psi}$ can be regarded as the free energy for the isothermal deformation $B_\Theta \to B$.

$$\psi\,(Y|\Theta,\tau) = \bar{\psi}(Y^{\Theta}|\Theta) = \frac{1}{\rho}\,W_{\Theta} + \phi(\Theta) \tag{1.12}$$

where ϕ is an arbitrary function of Θ and the Y^{Θ} are expressed by means of (1.9).

(1.12) allows to construct the function ψ when the form of he isothermal potential is known. The function ϕ can be determined when the reference configuration is free of stresses. It can be proved that

$$\phi = -\int_{\Theta_0}^{\Theta} \eta_{\xi}\, d\xi \tag{1.13}$$

where η_{ξ} is the specific entropy at the temperature ξ.

Analogous considerations can be developed starting from the isentropic potential.

2 - SOME REMARKABLE POTENTIALS.

Different forms of isothermal potentials have been proposed by various Authors, mostly on empirical basis. I shall quote here only few of them as examples.

The isothermal potential of Signorini's Second degree Elasticity is the following function of the inverse deformation tensor $\underline{e}$ (1) (Signorini[2])

$$W^{S} = \frac{1}{J}\{(\mu_{\tau} + \frac{C_{\tau}}{2})(I_{\underline{e}}+1) + \frac{1}{2}(\lambda_{\tau}+\mu_{\tau} - \frac{C_{\tau}}{2})I_{\underline{e}}^{2} + C_{\tau}II_{\underline{e}}\} - (\mu_{\tau} + \frac{C_{\tau}}{2}), \tag{2.1}$$

(1) The inverse deformation tensor $\underline{e}$ is defined by [see Ch. I n. 1]

$$\underline{1} + 2\underline{e} = (\underline{F}^{T})^{-1}\underline{F}^{-1} = \underline{B}^{-1} = \underline{c} = \underline{R}^{T}\,\underline{C}^{-1}\,\underline{R}\,.$$

where λ_τ and μ_τ are constants subject to the conditions

$$\mu_\tau > 0 \qquad 9\lambda_\tau + 5\mu_\tau > 0 \ . \tag{2.2}$$

These conditions are necessary and sufficient for the stability (if $c_\tau = 0$) of the reference configuration in the sense that W is positive definite and the quadratic form obtained from (2.1) as its first approximation is itself definite positive. $I_{\underline{e}}$ and $II_{\underline{e}}$ are the first and second invariants of the inverse deformation tensor, and $\bar{J} = \det \underline{F}^{-1} = 1/J$.

The form of Signorini's potential was suggested by the hypothesis that the principal components of the stress are polynomial functions of the second degree of the principal components of the tensor $\underline{e}$. Of course, for infinitesimal deformations W^S reduces to the classical Hookean potential

$$W = \frac{1}{2}(\lambda_\tau + 2\mu_\tau)\, I_{\underline{e}}^2 - 2\mu II_{\underline{e}} \tag{2.3}$$

The free energy corresponding to the isothermal potential of Signorini is given by

$$\psi^S = \frac{1}{\rho_o \bar{J}}\{-p + m(I_{\underline{e}}+1) + nI_{\underline{e}}^2 + cII_{\underline{e}}\} - q \tag{2.4}$$

where p, q, m, n, c are function of Θ and τ. Using Tolotti's theorem, we can express these last functions as function of λ and μ. For instance

$$p = \frac{\ell^2 - \ell_\tau^2}{8\,\ell_\tau^4}\{(\lambda+\mu)\ell^2 - (9\lambda_\tau + 5\mu_\tau)\,\ell_\tau^2\} \tag{2.5}$$

where λ and μ are the coefficients at the temperature Θ and λ_τ, μ_τ the same coefficients at the temperature τ.

For an incompressible material such that

$$J = f(\Theta,\tau) \quad , \quad f(\tau,\tau) = 1 \quad , \tag{2.6}$$

Signorini (Signorini[3]) introduced the potential

$$2\,W = c_1\,(I_{\underline{c}} - 3) + c_2(II_{\underline{c}}-3) + c_3(I_{\underline{c}}-3)^2 \qquad (2.7)$$

with c_1, c_2, c_3 functions of τ, wich reduces, for $c_3 = 0$ to the potential proposed by Mooney and extensively investigated by Rivlin (2).

Tolotti investigated the possibility of the propagation of acceleration waves in the case of axially symmetric Fresnel ellipsoid the symmetry axis coinciding with the propagation direction (Tolotti[6]) (3). He proved that a necessary and sufficient condition for the acceleration wave to be possible is that the potential have the form

$$W^T = \phi(J,\tau) + a(\tau)\, I_{\underline{E}} \qquad (2.8)$$

with $a > 0$ and ϕ satisfying some qualitative conditions. The corresponding form of the free energy is

---

(2) The conditions necessary and sufficient for W to be positive definite are very complicated (Manacorda[4]). It can be proved that these conditions are not sufficient to assure propagation of acceleration waves. This provides a counterexample to the belief that positive definitess of the potential always assures the possibility of wave propagation (Manacorda[5]).

(3) The condition for the propagation of an acceleration wave in the direction $\underline{N}$ to be possible is given by

$$(\underline{Q} - U_N^2\,\underline{1})\,\underline{A} = \underline{0} \qquad \text{or} \qquad (Q_{ij} - U_N^2\,\delta_{ij})\,A^j = 0$$

where $\underline{A}$ is the amplitude of the wave, U_N its speed of propagation and $\underline{Q}$ a tensor function of $\underline{N}$. In the present case, $\underline{Q}$ is symmetric, $\underline{Q}=\underline{Q}^T$, so that a characteristic quadric can be introduced whose equation is $\underline{Q}\,\underline{\lambda}\cdot\underline{\lambda}$ = const. If $\underline{Q}$ is positive definite, this quadric is an ellipsoid which is called Fresnel or polarisation ellipsoid.

$$\psi^T = \frac{1}{\rho}\phi\,(\Theta,\frac{\rho}{\rho_\tau}\,J) + \frac{a(\Theta)}{\rho}\{\frac{\ell_\tau^2}{\ell^2}(\frac{3}{2} + I_{\underline{E}}) - \frac{3}{2}\} + \beta(\Theta). \qquad (2.9)$$

Some different form of the elastic potential, which include as a particular case the potential proposed by Tolotti, have been proposed by Grioli (Grioli[6]) and, quite recently, by Boillat and Ruggeri in a paper on the propagation of unidimensional waves.

In two papers on materials endowed with rubber-like behaviour, Chadwick (Chadwick[9]) assumed a free energy of the form

$$\psi^C = \frac{\mu_\tau}{\rho_\tau} f(I_{\underline{E}},II_{\underline{E}},J;\tau)\frac{\Theta}{\tau} + \frac{k_\tau}{\rho_\tau}\{g(J,\tau)-\alpha_\tau h(J,\tau)(\Theta-\tau)\} - \int_\tau^\Theta (\frac{\Theta}{\xi} - 1)\, c_v\, d\xi \qquad (2.10)$$

where μ_τ, k_τ and α_τ are material constante, c_v is the specific heat at constant deformation (in Chadwick's theory, a function of temperature only), and f, g, h and α obey some qualitative conditions.

The corresponding isothermal potential is

$$W^C = \mu_\tau\, f(I_{\underline{E}},II_{\underline{E}},J;\tau) + k_\tau\, g(J,\tau) \qquad (2.11)$$

because, as a consequence of the conditions over f and g, $f(0,0,1;\tau)$ and $g(1;\tau)$ must vanish.

R E F E R E N C E S

1. Tolotti, C., Sul potenziale termodinamico dei solidi elastici omogenei ed isotropi per trasformazioni finite, *Atti R. Accad. Italia,* 14, 529, 1943.

2. Signorini, A., Trasformazioni termoelastiche finite, Mem. II, *Ann. Mat. pura Appl.* (4), 30, 1, 1949.

3. Signorini, A., Trasformazioni termoelastiche finite, Mem. III, Solidi incomprimibili, *Ann. Mat. pura appl.* (4), 39, 147, 1955.

4. Manacorda, T., Sul potenziale isotermo nella più generale elasticità di secondo grado per solidi incomprimibili. *Ann. Mat. pura appl.* (4), 40, 77, 1956.

5. Manacorda, T., On the waves propagation in Signorini's incompressible materials, *Boll. U.M.I.* (4), 5, 234, 1972.

6. Tolotti, C., Deformazioni elastiche finite, onde ordinarie di discontinuità e caso tipico dei solidi isotropi, *Rend. Mat. appl.* (5), 4, 33, 1943.

7. Grioli, G., On the thermodynamic potential for continuums with reversible deformations. Some possible types. *Meccanica,* 1, 15, 1966.

8. Boillat, G. & Ruggeri, T., Su alcune classi di potenziali termodinamici come conseguenza dell'esistenza di particolari onde di discontinuità nella meccanica dei continui con deformazioni finite, *Rend. Sem. Mat. Padova,*51,1,1974.

9. Chadwick, P., Thermo-mechanics of rubber-like materials. *Phil. Trans. Roy. Soc. Lond.,* 276, 371, 1974.

CHAPTER IV

LINEARIZATION. SIGNORINI'S PROBLEM

1 - LINEARIZATION.

Let $\underline{u}$ be the displacement of a continuous body B

$$\underline{u} = \underline{x} - \underline{X} \quad , \tag{1.1}$$

and let $\underline{u}$ be an analytic function of a parameter ε. The function

$$\bar{\underline{u}} = \left.\frac{d\underline{u}}{d\varepsilon}\right|_{\varepsilon=0} \tag{1.2}$$

is the coefficient of the first term in the development of $\underline{u}$ in a McLaurin series; $\bar{\underline{u}}$ will be named the linearized displacement of the body. Due to the formulae:

$$\underline{H} = \text{grad } \underline{u} = \underline{F} - \underline{1} \tag{1.3}$$

it can be obtained successively

$$\bar{\underline{H}} = \text{grad } \bar{\underline{u}} = \bar{\underline{F}} \tag{1.4}$$

and

$$\begin{aligned} \bar{\underline{C}} &= \overline{(\underline{F}^T \underline{F})} = \bar{\underline{F}}^T + \bar{\underline{F}} = \bar{\underline{H}}^T + \bar{\underline{H}} \quad , \\ \bar{\underline{E}} &= \frac{1}{2} (\bar{\underline{H}}^T + \bar{\underline{H}}) \end{aligned} \tag{1.5}$$

(in components: $E_{HK} = \frac{1}{2} (u_{U,K} + u_{K,H})$). As

$$\underline{F} = \underline{R}\, U \quad , \quad \underline{U}^2 = \underline{C} \quad ,$$

we also obtain

$$\bar{\underline{R}} = \frac{1}{2} (\bar{\underline{H}} - \bar{\underline{H}}^T) \ . \qquad (1.6)$$

The meaning of (1.4), (1.5), (1.6) is obvious. For istance,

$$\underline{R}\ (\varepsilon) = \underline{1} + \bar{\underline{R}}\ \varepsilon + O(\varepsilon^2) \qquad (1.7)$$

From (1.5) and (1.6) one obtains also

$$\bar{\underline{H}} = \bar{\underline{R}} + \bar{\underline{E}} \ . \qquad (1.8)$$

Coming back to (1.19) of the Ch. II we have

$$\underline{T}_o = \rho_o\ \partial_{\underline{E}}\psi \qquad \underline{T}_o = J\ \underline{F}^{-1}\ \underline{T}\ (\underline{F}^T)^{-1} \ ,$$

and assuming that the reference configuration is stress free, we obtain the linearized stress-strain law

$$\bar{\underline{T}}_o = \bar{\underline{T}} = \rho_o \frac{d}{d\varepsilon}\, \partial_{\underline{E}}\psi \Big|_{\varepsilon=0} = \rho_o\ \underline{L}[\bar{\underline{E}}] + \rho_o\ \underline{\alpha}\ \bar{\Theta} \qquad (1.9)$$

where $\underline{L}$ is a fourth-order tensor, $\underline{\alpha}$ a second-order tensor and both $\underline{L}$ and $\underline{\alpha}$ are functions at most of $\underline{X}$ and $\underline{x}$. In the mechanical case, $\underline{\alpha} = \underline{0}$ and $\underline{L}$ is the stiffness tensor of the material [1]. For an isotropic material, (1.9) reduces to the usual

[1] $\underline{L}$ is a linear operator mapping of the space of the symmetric double tensor into itself. If we regard the symmetric double tensors as vectors in a six dimensional space V_6, $\underline{L}$ is a linear operator on V_6, and therefore has only 36 distinct components. However, in the present case, only 21 components are independent, due to the symmetries of (1.9), as is well known.

form

$$\underline{\bar{T}} = \lambda(\mathrm{tr}\ \underline{\bar{E}})\ \underline{1} + 2\ \mu\ \underline{\bar{E}}\quad . \tag{1.10}$$

We now can come back to Cauchy equations

$$\rho_o\ \underline{b} + \mathrm{div}\ \underline{T}_R - \rho_o\ \underline{a} = \underline{0}\ ,$$

$$\underline{T}_R\ \underline{N} = \underline{f}_R ,$$

where $\underline{T}_R = J\ \underline{T}(\underline{F}^T)^{-1}$; we shall assume that $\underline{b}$ and $\underline{f}_R$ have the form

$$\underline{b} = \varepsilon\ \underline{\bar{b}}\quad ,\quad \underline{f}_R = \varepsilon\ \underline{f}\quad . \tag{1.11}$$

Linearization of the Cauchy equations gives therefore the equations

$$\rho_o\ \underline{\bar{b}} + \mathrm{div}\ \underline{\bar{T}} - \rho_o\ \underline{a} = 0$$

$$\underline{\bar{T}}\ \underline{N} = \underline{f}$$

where $\underline{\bar{T}}$ is given by (1.9) (with $\underline{\alpha} = 0$). The formulation(1.12) does not differ from the classical one of linearized elasticity.

2 - SMALL DEFORMATIONS SUPERIMPOSED ON A FINITE DEFORMATION.

Let B_o be a reference configuration of the body $\mathcal{B}$, and let B_1 be a configuration obtained from B_o, by a finite deformation whose gradient is $\underline{F}_o$; B_1 does not need to be an equilibrium configuration of $\mathcal{B}$. In any case, I shall write $\underline{T}_F^o$ for the Piola-Kirchhoff stress tensor of the deformation $B_o \to B_1$

and $\underline{T}_1$ for the Cauchy stress tensor calculated in B_1.

Moreover, I shall adopt the notations:

$\underline{F}$: gradient of deformation for $B_1 \rightarrow B$;

$\underline{F}^*$: gradient of deformation for $B_o \rightarrow B$;

$\underline{T}_R$: Piola-Kirchhoff stress tensor for the deformation $B_1 \rightarrow B$.

The position of a particle in B_o is denoted by $\underline{X}$, its position in B_1 by $\underline{x}$, and its position in B by $\underline{x}^*$. We shall consider infinitesimal deformations $B_1 \rightarrow B$ and ask for the Cauchy stress tensor $\underline{T}$ in B.

Of course .

$$\underline{F}^* = \underline{F}\,\underline{F}_o = (\underline{1} + \underline{H})\,\underline{F}_o \quad , \tag{2.1}$$

$$\underline{T}_1 = J^{-1}\,\underline{T}_R^o\,\underline{F}^T \quad , \tag{2.2}$$

$$\underline{T} = J_*^{-1}\,\underline{T}_R^*\,\underline{F}_*^T \quad , \tag{2.3}$$

where

$$J_o = \det \underline{F}_o \quad , \quad J^* = \det \underline{F}^* \tag{2.4}$$

Now, if the body is elastic, $\underline{T}_R$ is a function of $\underline{F}$ ($\underline{T}_R = \rho_o\,\partial_{\underline{F}}\psi$) $\underline{T}_R = \underline{f}(\underline{F})$; therefore, we obtain

$$\underline{T}_1 = J_o^{-1}\underline{f}_o(\underline{F}_o)\,\underline{F}_o^T = \frac{\rho_1}{\rho_o}\,\underline{f}_o(\underline{F}_o)\,\underline{F}_o^T \qquad (\frac{\rho_1}{\rho_o} = J_o^{-1}) \tag{2.5}$$

where $\underline{f}_o$ is the response function relative to the reference configuration B_o, and

$$\underline{T}_R^* = \underline{f}_o(\underline{F}^*) = \underline{f}_o((\underline{1} + \underline{H})\,\underline{F}_o) \tag{2.6}$$

or linearizing,

$$\bar{\underline{T}}^*_R = \underline{f}_o\ (\underline{F}_o) + \underline{A}_o\ [\bar{\underline{H}}\ \underline{F}_o] \tag{2.7}$$

where

$$\underline{A}_o = d_{\underline{F}^*}\ \underline{f}_o \Big|_{\underline{F}^*=\underline{F}_o} \tag{2.8}$$

is the fourth-order stiffness tensor of the material calculated in B_1.

From (2.6) and (2.3) one finally obtains

$$\underline{T} = \frac{\rho}{\rho_o} \underline{T}^*_R\ \underline{F}^T_* = \frac{\rho}{\rho_o}\{\underline{f}_o(\underline{F}_o)+\underline{A}_o[\bar{\underline{H}}\ \bar{\underline{F}}_o]+0(2)\}\underline{F}^T_o(\underline{1}+\underline{H}^T)\ ,$$

$$\bar{\underline{T}} = \frac{\bar{\rho}}{\rho_1}\{\underline{T}_1 + \frac{\rho}{\rho_o}\ \underline{A}_o[\bar{\underline{H}}\ \underline{F}_o]\underline{F}^T_o\}(\underline{1}+\bar{\underline{H}}^T) = \tag{2.9}$$

$$= (1\text{-trac}\ \underline{E})\underline{T}_1(1+\bar{\underline{H}}^T)+ \frac{\rho_1}{\rho_o}\ \underline{A}_o[\bar{\underline{H}}\ \underline{F}_o]\underline{F}^T_o$$

a formule which defines $\underline{T}$ as a function of the linearized deformation $\bar{\underline{H}}$. In a more explicit form:

$$\bar{\underline{T}} = \underline{T}_1-(\text{tr}\ \bar{\underline{E}})\underline{T}_1+\underline{T}_1\bar{\underline{H}}^T + \frac{\rho_1}{\rho_o}\ \underline{A}_o\ [\bar{\underline{H}}\ \underline{F}_o]\ \underline{F}^T_o\ . \tag{2.10}$$

Now, for the Piola-Kirchhoff stress tensor $\underline{T}_R$ we have

$$\underline{T}_R = \frac{\rho_1}{\rho}\ \underline{T}\ (\underline{F}^T)^{-1} = \frac{\rho_1}{\rho}\ \frac{\rho}{\rho_o}\ \underline{T}^*_R\ \underline{F}^T_*\ (\underline{F}^T)^{-1}$$

that is

$$\bar{\underline{T}}_R = \frac{\rho_1}{\rho_o}\{\ \underline{f}_o(\underline{F}_o) + \underline{A}_o[\bar{\underline{H}}\ \underline{F}_o]\ \}\ \underline{F}^T_o = \underline{T}_1+ \frac{\rho_1}{\rho_o}\ \underline{A}_o[\bar{\underline{H}}\ \underline{F}_o]\underline{F}^T_o\ , \tag{2.11}$$

so that the equation of motion starting from B_1, $\text{div}_{\underline{x}}\ \underline{T}_R\ +$ $+\ \rho_1\underline{b} = \rho_1\ \ddot{\underline{x}}$, reduces to the linearized equation

$$\operatorname{div}_{\underline{x}} \underline{T}_1 + \operatorname{div}_{\underline{x}} \frac{\rho_1}{\rho_o} A_o [\bar{\underline{H}}\ \underline{F}_o]\ \underline{F}_o^T + \rho_1\ \underline{b} = \rho_1\ \ddot{\underline{x}}^* . \tag{2.12}$$

In particular, if B_1 is an equilibrium configuration under the action of the forces $\rho_1 \underline{b}$, and $\underline{b}$ does not depend on the motion, equation (2.12) reduces to

$$\operatorname{div}_{\underline{x}} \frac{\rho_1}{\rho_o} \underline{A}_o [\bar{\underline{H}}\ \underline{F}_o]\ \underline{F}_o^T = \rho_1\ \ddot{\underline{x}}^* . \tag{2.13}$$

The argument of the divergence is a linear function of $\bar{\underline{H}}$; we can therefore write

$$\underline{A}_o [\bar{\underline{H}}\ \underline{F}_o]\ \underline{F}_o^T = \underline{B}\ [\bar{\underline{H}}] \tag{2.14}$$

where $\underline{B}$ is once a more a fourth-order tensor depending of $\underline{x}$. Therefore (2.13) can be put in the final form

$$\operatorname{div}_{\underline{x}}\ \underline{B}\ [\bar{\underline{H}}] = \rho_1\ \ddot{\underline{x}}^* . \tag{2.15}$$

In the more general case, $\underline{B}$ is a function (of $\underline{X}$ and) of $\underline{x}$, so that, a body homogeneous in B_o loses its homogeneity in B_1. However, if the body is homogeneous in B_o and the deformation $B_o \to B_1$ is homogeneous, $\underline{B}$ is no longer a function of $\underline{x}$, and the equation of motion can be written (in components):

$$B_{ijhm}\ \bar{u}_{h,mj} = \rho_1\ \ddot{\bar{u}}_i \tag{2.16}$$

if $\underline{u} = \underline{x}^* - \underline{x}$ and if B_1 is an equilibrium configuration of the body.

As an example, we may ask for a solution of (2.16) of the form

$$\bar{\underline{u}} = \underline{a}\ \exp.\ i(\underline{k}\cdot \underline{x} - \omega t) \tag{2.17}$$

where $\underline{a}$ and $\underline{k}$ are constant vectors (plane infinitesimal waves).

We obtain $\bar{u}_{hmj} = -\bar{u}_h k_m k_j$, $\ddot{\bar{\underline{u}}} = -\omega^2 \bar{\underline{u}}$, so that the equation of motion is equivalent to

$$B_{ijhm} k_m k_j a_h = \rho_1 \omega^2 a_i$$

Putting

$$\underline{n} = \underline{k} / |\underline{k}| \quad , \quad Q_{ih} = B_{ijhm} n_j n_m$$

we obtain finally

$$(\underline{Q} - \rho_1 U^2 \underline{1}) \underline{a} = 0 \tag{2.18}$$

where $U = \omega^2 / \underline{k}^2$.

If $\underline{Q}$ is symmetric, it has three mutually orthogonal proper directions, and three real proper numbers.

(2.18) proves that, for any $\underline{n}$, $\underline{a}$ must be a proper direction of $\underline{Q}$ and $\rho_1 U^2$ a proper number. Indeed, for (2.17) have a solution, the det $(\underline{Q} - \rho_1 U^2 \underline{1})$ must vanish. The proper numbers (divided by ρ_1) are the speeds of propagation in the direction of the proper vectors of $\underline{Q}$ only if they are positive, that is, only if the quadratic form $\underline{Q}\, \underline{\nu} \cdot \underline{\nu}$ is positive definite or if the expression

$$B_{ijhm} \lambda_j \lambda_m \nu_i \nu_h \tag{2.19}$$

is positive for all $\underline{\lambda}$ and $\underline{\nu}$. It followa that if the fourth-order tensor $\underline{B}$ is strongly elliptic for any direction of propagation $\underline{n}$, there are three mutually orthogonal directions of propagation and three speeds of propagation.

3 - THE SOLUTION OF PROBLEMS OF ELASTOSTATICS BY SUCCESSIVE APPROXIMATIONS.

Signorini (Signorini[1]) systematically developed a method of successive approximations in his papers on non linear elastostatics (1). The balance equations of elastostatics are:

$$\begin{aligned} \operatorname{div} \underline{T}_R + \rho_o \underline{b}_R &= 0 \qquad \text{in } B_o \;, \\ \underline{T}_R \underline{N} &= \underline{f}_R \qquad \text{on } \partial B_o \;. \end{aligned} \tag{3.1}$$

If we put further, $\underline{b}_R = \varepsilon \underline{b}$, $\underline{f}_R = \varepsilon \underline{f}$ (2), where $\underline{b}$ and $\underline{f}$ are given functions of the points of B_o and ∂B_o respectively, we can lock for a solution of the form

$$\underline{u} = \sum^{\infty} \varepsilon^n \underline{u}_n \;, \tag{3.2}$$

where $\underline{u}_1 = \bar{\underline{u}}$ is the "linearization" of $\underline{u}$.

It is an easy matter to show that a necessary condition for (3.1) to have a solution is that the applied forces be equilibrated, i.e.

$$\int_{B_o} \rho_o \underline{b}_R \, dV + \int_{\partial B_o} \underline{f}_R \, d\Sigma = \underline{0} \;, \tag{3.3}$$

(1) Signorini's method can be easily adapted to non-linear elastodynamics.

(2) More generally, a decomposition of the form: $\underline{b}_R = \varepsilon \underline{b}_1 + B(\varepsilon)$, $\underline{f}_R = \varepsilon \underline{f}_1 + F(\varepsilon)$ is usually assumed in the modern treatments of the subject.

and

$$\int_{B_o} \underline{p} \times \rho_o \underline{b}_R \, dV + \int_{B_o} \underline{p} \times \underline{f}_R \, d\Sigma = \underline{0} \,, \tag{3.4}$$

where $\underline{p} = \underline{x} - \underline{X}_o$ is the position vector of $\underline{x}$ with respect to a fixed point $\underline{X}_o$. As a consequence of Da Silva's theorem (3) (Da Silva[4]), we shall always assume that the forces are equilibrated in the reference configuration, i.e. that (3.3) and

$$\int_{B_o} \underline{p}_R \times \rho_o \underline{b}_R \, dV + \int_{\partial B_o} \underline{p}_R \times \underline{f}_R \, d\Sigma = \underline{0} \tag{3.5}$$

hold in the reference configuration.

As $\underline{p} = \underline{X} - \underline{X}_o + \underline{x} - \underline{X} = \underline{p}_R + \underline{u}$, (3.4) results in a restriction for the admissible successive displacements $\underline{u}_n$.

Let $\bar{\underline{u}}_1$ be a particular solution of the linearized equations (1.11); then the general solution is given by

$$\underline{u}_1 = \bar{\underline{u}}_1 + \underline{\omega}_1 \times (\underline{X}-\underline{X}_o) = \bar{\underline{u}}_1 + \underline{\omega}_1 \times \underline{p}_R \tag{3.6}$$

where $\underline{\omega}_1$ is a constant vector. Now (3.4) yields

$$\int_{B_o} \underline{u}_1 \times \rho_o \underline{b} \, dV + \int_{\partial B_o} \underline{u}_1 \times \underline{f} \, d\Sigma = \underline{0}$$

or

$$\int_{B_o} [\underline{\omega}_1 \times (\underline{X}-\underline{X}_o)] \times \rho_o \underline{b} \, dV + \int_{\partial B_o} [\underline{\omega}_1 \times (\underline{X}-\underline{X}_o)] \times \underline{f} \, d\Sigma = - \int_{B_o} \bar{\underline{u}}_1 \times \rho_o \underline{b} \, dV - \int_{\partial B_o} \bar{\underline{u}}_1 \times \underline{f} \, d\Sigma \tag{3.7}$$

---

(3) Let (3.3) hold: then there are at least four rigid rotations of the body such that (3.5) holds.

which results in an equation for the unknown infinitesimal rotation $\underline{\omega}_1$.

If we introduce the static load tensor

$$\underline{A} = \int_{B_o} \underline{p} \otimes \rho_o \underline{b} \, dV + \int_{\partial B_o} \underline{p} \otimes \underline{f} \, d\Sigma \tag{3.8}$$

and we put $\underline{C} = \underline{A} - (\mathrm{tr}\, \underline{A})\, \underline{1}$, we can write (2.7) in the form

$$\begin{aligned} \underline{C}_R \, \underline{\omega}_1 &= (\underline{A}_R - \mathrm{tr}\, \underline{A}_R \, \underline{1})\, \underline{\omega}_1 = - \underline{M} \\ \underline{M}_1 &= \int_{B_o} \bar{\underline{u}}_1 \times \rho_o \underline{b} \, dV + \int_{\partial B_o} \underline{u}_1 \times \underline{f} \, d\Sigma \end{aligned} \tag{3.9}$$

where $\underline{A}_R$ is the static loads tensor in the reference configuration. If det $\underline{C}_R \neq 0$, (3.9) can be solved for $\underline{\omega}_1$ which, therefore, is no longer arbitrary as is the case in linear elastostatics. But, if det $\underline{C}_R = 0$, i.e. if tr $\underline{A}_R$ is a proper number of $\underline{A}_R$, (3.9) can admit infinitely many solutions, or even no solution at all.

In this last case, the linearized problem is incompatible (in the sense of Signorini) with the non-linear one.

4 - THE SUCCESSIVE APPROXIMATIONS.

The equations of the n-th approximation are obtained by differentiating n times the non-linear equations (3.1) with respect to ε and then putting ε = 0; these equations read:

$$\begin{aligned} \mathrm{div}\, \underline{L}\, [\underline{E}_n] + \rho_o \underline{b}_n &= 0 \\ \underline{L}\, [\underline{E}_n]\, \underline{N} &= \underline{f}_n \end{aligned} \tag{4.1}$$

where $\underline{E}_n$ is the linearized deformation tensor of the n.th displacement and $\underline{b}_n$, $\underline{f}_n$ are known functions of the first, second, ..., (n-1)-th approximations $\underline{u}_k$.

If the n-1 displacements $\underline{u}_1, \ldots, \underline{u}_{n-1}$ have been already determined (as well as the rotations $\underline{\omega}_1, \ldots \underline{\omega}_{n-1}$); then the n-th rotation will be determined by the equation

$$\underline{C}_R \, \underline{\omega}_n = - \underline{M}_n \tag{4.2}$$

Therefore, it is clear that $\underline{\omega}_n$ can be univocally determined if and only if det $\underline{C}_R \neq 0$. In the complementary instance, i. e. when det $C_R = 0$, en incompatibility of the order n emerges, which can render the classical method meaningless (1).

---

(1) It can be proved that (4.2) are the condition of integrability of the (n-1)-th system. (Signorini[1]).

5 - THE DYNAMICAL EXPLANATION.

Incompatibilities can occur only if the system of loads admits an axis of equilibrium [1]. Let the first, the second, ..., the (n-1)-th conditions be satisfied, but not the n-th; Capriz suggested that this case can occur when the system of equations admits a dynamical solution for which the acceleration is infinitesimal of the order n with respect to the parameter ε, so that the first n-1 equations can be satisfied by statical solution. Quite recently, Capriz and Podio Guidugli gave to this ides an appropriate form in a very exaustive paper [2] on this subject (Capriz & Podio Guidugli[5]). By exploiting this definition, they were able to prove that incompatibilities do not occur if and only if the system of loads is infinitesimally stable.

---

(1) Let $(\underline{X}_o, r)$ be a line issuing from the point $\underline{X}_o$, and let $\underline{e}$ be a unit vector parallel to r, further, let the system of loads $\{\rho_o\underline{b}, \underline{f}\}$ be equilibrated with respect to $\underline{X}_o$; then r is an axis of equilibrium if all rotations about $(\underline{X}_o, r)$ that do not violate the equilibrium of the system of loads, i.e. if the equation

$$\int_{B_o} \underline{Q}\,(\underline{X} - \underline{X}_o) \times \rho_o\underline{b}\; dV + \int_{\partial B_o} \underline{Q}(\underline{X} - \underline{X}_o) \times \underline{f}\; d\Sigma = \underline{0}$$

holds for all orthogonal $\underline{Q}$ such that $\underline{Q}\,\underline{e} = \underline{e}$. This condition must therefore be satisfied also by infinitesimal rotations, for which $\underline{Q} = \underline{\omega}\times$. The condition for an axis to be equilibrium axis is therefore that the equation

$$(*)\quad \int_{B_o} [\underline{\omega}\times(\underline{X} - \underline{X}_o)] \times \rho_o\underline{b}\; dV + \int_{\partial B_o} [\underline{\omega}\times(\underline{X} - \underline{X}_o)] \times \underline{f}\; d\Sigma = \underline{0}$$

hold for all $\underline{\omega}$ parallel to $\underline{e}$. Equation (X) can be written as $\underline{C}\,\underline{\omega} = 0$; thus, $\underline{e}$ can be axis of equilibrium if and only if $\underline{C}$ is singular, and $\underline{e}$ is a proper vector of $\underline{A}$.

(2) In which, referring to previous papers by Capriz, a statical interpretation is also given.

Let λ be a real parameter and put

$$\underline{A}(\lambda) = \underline{A}_R + \lambda \left\{ \int_{B_o} \underline{u} \otimes \rho_o \underline{b} \, dV + \int_{\partial B_o} \underline{u} \otimes \underline{f} \, d\Sigma \right\} \quad ; \qquad (4.1)$$

the system of loads is called infinitesimally stable if, for any displacement $\underline{u}$, a neighborhood Λ of $\lambda = 0$ exist such that the system of loads is equilibrated for all orthogonal continuously differentiable $\underline{Q}(\lambda)$ endowed with the property that $\underline{Q}(\lambda) - \underline{1}$ tends to zero in norm when λ tends to zero(3). This fundamental theorem proves that a sufficient condition for the method of successive approximation to be consistent is that the system of loads be infinitesimally stable, a property of the reference configuration only.

(3) A system of loads which does not have an axis of equilibrium is infinitesimally stable.

REFERENCES

1. Signorini, A., Trasformazioni termoelastiche finite, Mem. II, *Ann. Mat. pura appl.* (4), 30, 1, 1949.

2. Signorini, A., Un semplice esempio di "incompatibilità" tra la elastostatica classica e la teoria delle deformazioni elastiche finite, *Acc. Naz. Lincei, Rend. Cl. Fis. Mat. Nat.* (8), 8, 276, 1950.

3. Tolotti, C., Orientamenti principali di un corpo elastico rispetto alla sua sollecitazione totale, *Mem. Acc. It.* (7), 13, 1139, 1943.

4. Da Silva, D.A., Memoria sobre a rotação das forças em torno das pontos d'applicação, *Mem. Ac. R. Sci. Lisboa* (2), 2, (I), 61, 1851.

5. Capriz, G., & Podio Guidugli, P., On Signorini's perturbation method in finite elasticity, *Arch. Rat. Mech. An.*, 57, 1, 1974.

6. Capriz. G., Sopra le deformazioni elastiche finite di un solido tubolare, *Rend. di Mat.* (5), 15, 228, 1956.

7. Capriz, G., Sui casi di "incompatibilità" tra l'elastostatica classica e la teoria delle deformazioni elastiche finite, *Riv. Mat. Univ. Parma,* 10, 119, 1959.

8. Capriz, G., A remark on Signorini's "phenomena of incompatibility", *Bul. Inst. Pol. Iasi,* 11 (XV), 307, 1965.

9. Zorski, H., On the equations describing small deformations superposed on finite deformation, in *Proc. Intern. Symp. Second-order Effects,* Haifa, 1962, 109.

BIBLIOGRAPHY

1. Eringen, C.A., & Suhubi, E., *Non-linear elastodynamics*, Academic Press, New York, 1974.
2. Grioli, G., *Mathematical theory of elastic equilibrium, (Recent results)*, Springer, Berlin, 1962.
3. Müller, I., *Thermodynamik Die Grundlagen der Matherialtheorie*, Bartelsmann, Dortmund, 1973.
4. Nowacki, W., *Thermoelasticity*, Pergamon Press, Oxford, 1962.
5. Signorini, A., Trasformazioni termoelastiche finite, Mem. I, *Ann. Mat. pura appl.* (4), 27, 33, 1943; Mem. II, *Ann. Mat. pura appl.* (4), 30, 1, 1949; Mem. III, *Ann. Mat. pura appl.* (4), 39, 147, 1955; Mem. IV, *Ann. Mat. pura appl.* (4), 51, 329, 1960.
6. Truesdell, C., *Rational thermodynamics*, McGraw Hill, New York, 1969.
7. Truesdell, C., *Introduction à la mécanique rationelle des milieux continus*, Masson, Paris, 1974.
8. Truesdell, C., & Toupin, R., The classical field theories, in *Encyclopedia of Physics*, Vol. III/1, Springer, Berlin, 1960.
9. Truesdell, C., & Noll, W., The non-linear field theories of mechanics, in *Encyclopedia of Physics*, Vol. III/3, Springer, Berlin, 1965.
10. Truesdell, C., & Wang, C.C., *Introduction to rational elasticity*, Nordhoof, Leyden, 1973.

ANALYTICAL MECHANICS OF ELASTIC MEDIA

Czeslaw WOZNIAK
University of Warsaw, Poland
03-922 Warszawa, Miedzynarodowa 58 m.63

1. Introduction

Using the term "analytical mechanics" we usually mean the specific kind of presentation of mechanics of the finite systems of particles; the concept of constraints and that of the generalized coordinates plays an important role in this presentation. In this paper we are to deal with mechanics of uncountable systems of particles constituting material continua. However, our way of presentation is different from that used in the classical approach to continuum mechanics. We start from the dynamics of an uncountable system of homogenously deformable elastic particles and we impose certain restrictions on the motion or on the internal forces. Such system has much more general structure then the well known classical elastic continuum being characterized not only by its material properties but also by the restrictions imposed on the motion and the system of internal forces. It can be observed that all known "technical" or "approximate" theories of elasticity (theories of elastic plates, shells, rods as well as different finite element approaches, formulations based on the Ritz or Galerkin methods, etc.) can be directly derived form the analytical mechanics of elastic media presented in this contribution. At the same time, mechanics of such "structured" media is also of the more general nature than the mechanics of constrained continuum [1,2], where only restrictions for the deformations are taken into account.

The results given in the contribution are based on [3, 4] ; for more detailed analysis and applications of the theory the reader is referred to [5-7] .

2. Continuum of Particles : Primitive Concepts, Laws of Dynamics and Definition of External Loads.

To formulate the axioms of mechanics we have to introduce first the primitive concepts on which our approach is based. We introduce the well known concept of space-time with a given inertial coordinates system x^k, t , where $x \equiv (x^k)$ are orthogonal Cartesian coordinates in the physical space and t is the time coordinate(*). We are to deal with a continuum of particles in which each particle will be distinguished by the position vector in certain fixed reference configuration $X \equiv (X^\alpha) \in B_R$, B_R being a region in the physical space and X^α are said to be material coordinates of the particle. Now we can introduce the second primitive concept, namely the motion of the continuum, which will be represented by the vector function $x = \chi(X,t)$, $X \in B_R, t \in R$ (the deformation function). We assume that $\chi = (\chi^k)$ is continuous with the first and second material derivatives $\nabla\chi \equiv (\chi^k{}_{,\alpha})$, $\nabla\nabla\chi \equiv (\chi^k{}_{,\alpha\beta})$, first and second time derivatives $\dot{\chi}$, $\ddot{\chi}$, and that the relation $\det \nabla\chi > 0$ holds for each $X \in B_R$, $t \in R$. The next primitive concept is that the mass, which will be represented by the mass density $\varrho(X,t), X \in B_R, t \in R$. We postulate the known mass conservation law represented by $\varrho = \varrho_R \mathcal{J}^{-1}$, $\mathcal{J} \equiv \det \nabla\chi$, where $\varrho_R(X)$, $X \in B_R$, is the mass density in the reference configuration. We introduce as the primitive concept the system of forces $\{b, d_R, p_R, t_R\}$, where $b = (b^k(X,t))$ is the external body load, $d_R = (d^k_R(X,t))$ is the internal interactions density (both fields are defined on $B_R \times R$), $p_R = (p^k_R(X,t))$ is a density of external surface tractions and $t_R = (t^k_R(X,t))$ is a density of internal surface tractions (these fields are defined nearly everywhere on $\partial B \times R$).

(*) The indices $\alpha, \beta, \gamma, \delta$ and k, l run over the sequence 1, 2, 3. The summation convention holds for all kinds of indices. The partial derivatives of function $f(X,t)$ with respect to X^α are denoted by $f_{,\alpha}$ and with respect to t by $\dot{f}$.

The index $\mathbf{R}$ informes us that the density under consideration is related to the reference configuration of the continuum ; if q_R and q are volume densities of the same quantity related to the reference and actual configuration, respectively, then $q = q_R \mathcal{J}^{-1}$, $\mathcal{J} \equiv \det \nabla\chi$. The system of forces $\{\mathbf{b}, \mathbf{d}_R, \mathbf{p}_R, \mathbf{t}_R\}$ and the motion of the continuum are assumed to be interrelated by means of the Newton's laws of dynamics ; we postulate that in $B_R \times R$ the following relation holds

$$\varrho_R \ddot{\chi} = \varrho_R \mathbf{b} + \mathbf{d}_R , \tag{1.1}$$

and on $\partial B_R \times R$ the principle of action and reaction gives

$$\mathbf{p}_R + \mathbf{t}_R = \mathbf{0} \tag{1.2}$$

The fields of forces are assumed to be continuous (nearly everywhere) in the domains of their definitions. From (1.1) and (1.2) we obtain the equivalent relations : $\varrho\ddot{\chi} = \varrho\mathbf{b} + \mathbf{d}$ in $B_R \times R$, and $\mathbf{p} + \mathbf{t} = \mathbf{0}$ on $\partial B_R \times R$.

The external body loads $\mathbf{b}$ and the density of external surface loads $\mathbf{p}_R$ are assumed to be known function of the motion in each problem under consideration. It means that in each special problem of mechanics there are given the following load relations

$$\mathbf{b} = \beta(\chi), \quad \mathbf{p}_R = \pi_R(\chi) , \tag{1.3}$$

where β and π_R are known operators. In the special case these operators can reduce to the known function of the material coordinates X^α only ; the external loads are then called the "dead" loads.

Mind that, while Eqs. (1.1) and (1.2) express the physical laws and hold for any continuum of material particles we deal with, the form of equations (1.3) has to be specified for any problem under consideration because the external loads in different problems can be different.

2. Elastic Particles. Free and Constrained Elastic Continua.

In our presentation we shall interpret the deformation function $\chi(\mathbf{X},\cdot)$, restricted to an arbitrary but fixed point $\mathbf{X}$, as the function describing the motion of a sufficiently small piece of the matter, $\mathbf{X}$ being the position vector of the mass center of this piece in the reference configuration. Each such piece of the matter we shall refer to as the particle.

To characterize the material properties of the particle we apply the well known local theory of the constitutive equations. To this aid we assign to each particle the local deformation given by the non-singular 3 x 3 matrix $\mathbf{F} \equiv \left(F_{k\alpha}(\mathbf{X}, t) \right)$ (for any $\mathbf{X} \in B_R$ and any time instant t) and the local state of stress (given for example by the first Piola–Kirchhoff stress tensor $\mathbf{T}_R = \left(T_R^{k\alpha}(\mathbf{X}, t) \right)$, $\mathbf{X} \in B_R$, $t \in R$). We shall confine ourselves to the hyperelastic materials only ; they are the materials in which the stress assigned to the particle is related to the local deformation by means of the relation

$$\mathbf{T}_R = \varrho_R \frac{\partial \sigma}{\partial \mathbf{F}} , \tag{2.1}$$

where $\sigma = \sigma(\mathbf{X}, t)$ is a strain energy function. The form of the strain energy function has to be invariant under arbitrary rigid rotations of the physical space ; such situation occurs only if the function has the form $\sigma = \hat{\sigma}(\mathbf{X}, \mathbf{C})$ where $\mathbf{C} \equiv \mathbf{F}^T \mathbf{F}$ is a metric deformation tensor (the right Cauchy-Green deformation tensor). Defining the convective stress tensor $\bar{\mathbf{T}} \equiv \left(\bar{T}^{\alpha\beta} \right)$ by means of the relation

$$\bar{\mathbf{T}} \equiv J^{-1} \mathbf{F}^{-1} \mathbf{T}_R , \quad \bar{T}^{\alpha\beta} \equiv J^{-1} \, T_R^{k\beta} \left(F^{-1} \right)^{\alpha}_{k} , \tag{2.2}$$

we can rewrite the constitutive relation (2.1) in the equivalent form

$$\bar{\mathbf{T}} = 2 \varrho \frac{\partial \sigma}{\partial \mathbf{C}} . \tag{2.3}$$

Equations (2.1) or (2.3) describe the elastic materials the particles of the body are made of.

Let us assume, for the time being, that there are no interactions in the system of particles under consideration ; it means that $\mathbf{t}_R = \mathbf{0}$, $\mathbf{d}_R = \mathbf{0}$ and $\mathbf{T}_R = \mathbf{0}$. From (1.1) it follows that each particle moves independently of any other particle and from (2.3) we conclude, at least for solids, that the metric deformation tensor $\mathbf{C}$ is constant in time. Such continuum of particles will be called free and it does not represent, from the point of view of applications, any interesting object of analysis. Thus we are to investigate only such continua of particles in which there exist interparticle interactions, i.e. when the fields $\mathbf{t}_R$, $\mathbf{d}_R$, $\mathbf{T}_R$, are not equal to zero. To construct mechanics of these continua we shall postulate certain relations between the kinematic χ, $\mathbf{F}$ and the kinetic $\mathbf{t}_R, \mathbf{d}_R, \mathbf{T}_R$ fields. All such relations will be called structural restrictions and the continuum of particles, governed by Eqs. (1.1)-(1.4) and by the well defined structural restrictions, is said to be the structured continuum. Because we can postulate many types of structural restrictions we can construct different structured material continua.

Mind that the concept of structural restrictions introduced here is of a more general nature then the concept of constraints imposed on deformations only where, as the starting point of consideration, the classical material continuum is taken into account [1, 2].

3. Material Continua with Kinematic Restrictions. Ideal Restrictions for the Kinematic Fields. Passage to the Classical Non-linear Elasticity.

Let us assume that on the kinematic fields χ, $\mathbf{F}$, which are supposed to be sufficiently smooth, are imposed restrictions described by the following system of partial differential equations

$$h_\nu(\mathbf{X}, t, \chi, \nabla\chi, \mathbf{F}, \psi, \nabla\psi) = 0, \quad \nu = 1, \ldots, N, \quad \mathbf{X} \in B_R \tag{3.1}$$

and by the boundary conditions of the form

$$R_\varrho(\mathbf{X}, t, \chi, \psi) = 0, \quad \varrho = 1, \ldots, s, \quad \mathbf{X} \in \partial B_R, \tag{3.2}$$

where h_ν, R_ϱ are known differentiable functions of all arguments and $\psi = (\psi_a(\mathbf{X}, t))$, $a = 1, \ldots, n < N$, is a unknown vector function defined on $\bar{B}_R \times R$. We do not treat ψ as a new primitive concept but assume that the definition of ψ (in terms of χ, $\mathbf{F}$) is included into Eqs. (3.1). We also assume that there exists at least one (χ, $\mathbf{F}$, ψ) satisfying (3.1), (3.2) and the suitable smoothness conditions and that Eqs. (3.1), (3.2) do not reduce to the definition of the vector ψ only, but that they express certain restrictions on the kinematic fields χ and $\mathbf{F}$. Eqs. (3.1), (3.2) are defining equations ; more general form of these constraints (in which functions on the left-hand side of (3.1), (3.2) can also depend on the time derivatives) can be taken into account.

Using the same reasoning as in the analytical mechanics of finite system of particles, we postulate that there exist a connection between the form of restrictions for the kinematic fields and the system of reaction forces which maintain these restrictions. In what follows we shall confine ourselves to the material system in which the kinematic restrictions (3.1), (3.2) are ideal, i.e. the relation

$$\oint_{\partial B_R} \mathbf{t}_R \cdot \delta\boldsymbol{\chi} \, ds_R + \int_{B_R} (\mathbf{d}_R \cdot \delta\boldsymbol{\chi} + \mathbf{T}_R \cdot \delta\mathbf{F}) \, dv_R = 0 \tag{3.3}$$

holds for any $\delta\boldsymbol{\chi} \equiv (\delta\chi^k)$, $\delta\mathbf{F} \equiv (\delta F_{k\alpha})$ such that $(\delta\boldsymbol{\chi}, \delta\mathbf{F}, \delta\boldsymbol{\psi})$ is a solution of the following system of linear partial differential equations in B_R

$$\frac{\partial h_\nu}{\partial \chi_k}\delta\chi_k + \frac{\partial h_\nu}{\partial \chi_{k,\alpha}}\delta\chi_{k,\alpha} + \frac{\partial h_\nu}{\partial F_{k\alpha}}\delta F_{k\alpha} + \frac{\partial h_\nu}{\partial \psi_a}\delta\psi_a + \frac{\partial h_\nu}{\partial \psi_{a,\alpha}}\delta\psi_{a,\alpha} = 0 , \tag{3.4}$$

and the boundary conditions defined nearly everywhere on ∂B_R

$$\frac{\partial R_\varrho}{\partial \chi_k}\delta\chi_k + \frac{\partial R_\varrho}{\partial \psi_a}\delta\psi_a = 0 , \tag{3.5}$$

and where Eqs. (3.4), (3.5) are defined for any regular $\boldsymbol{\chi}$, $\mathbf{F}$, $\boldsymbol{\psi}$ satisfying (3.1) and (3.2)(*).

We can easily observe that if there are no kinematic restrictions (i.e. $h_\nu \equiv 0$, $R_\varrho \equiv 0$), then $\delta\boldsymbol{\chi}, \delta\mathbf{F}$ are arbitrary continuous functions (the coefficients in Eqs. (3.4), (3.5) are equal to zero) and from (3.3), by virtue of the known Du Bois-Reymonde lemma, we obtain $\mathbf{t}_R = \mathbf{0}$, $\mathbf{d}_R = \mathbf{0}$, $\mathbf{T}_R = \mathbf{0}$. It follows that if the kinematic restrictions are absent then all internal forces disappear and we deal with the free continuum of particles.

Let us analyse some other special cases of the material continua with the ideal kinematic restrictions.

(*) For the physical interpretation of Eq. (3.3) cf. [5].

If the functions h_ν in Eqs. (3.1) are independent of the prescribed component $F_{k\alpha}$ of $\mathbf{F}$ (subscripts k, α being fixed here), then $\delta F_{k\alpha}$ is an arbitrary continuous function and from (3.3) we conclude that $T_R^{k\alpha} = 0$. The stress relation (2.1) for $T_R^{k\alpha}$ reduce to $\partial\sigma / \partial F_{k\alpha} = 0$, subscripts k, α being fixed, and the latter equation can be treated as an equation for $F_{k\alpha}$ (*). It implies that we arrive at what can be called the slender continuum, i.e. the material continuum which is unable to carry the component $T_R^{k\alpha}$ of the stress tensor. More complicated special cases of the slender continua can be also considered [6].

If the functions h_ν are independent of all components of $\mathbf{F}$, then $\delta\mathbf{F}$ is an arbitrary function and from (3.3) it follows that we have constructed the material continuum in which there are no stresses : $\mathbf{T}_R = \mathbf{0}$. Eqs. (2.1) reduce to $\partial\sigma/\partial\mathbf{F} = \mathbf{0}$ and the dynamics of such continuum is independent of the material properties of the particles. An example of this type of continua is given by a rigid body ; Eqs. (3.1) then reduce to $(\nabla\chi)^T\nabla\chi - \mathbf{1} = \mathbf{0}$.

If among Eqs. (3.1) there is the equation $\nabla\chi - \mathbf{F} = \mathbf{0}$ then applying to (3.3) the divergence theorem we obtain

$$\oint_{\partial B_R} \mathbf{s}_R \cdot \delta\chi \, ds_R + \int_{B_R} \mathbf{r}_R \cdot \delta\chi \, dv_R = 0 , \tag{3.6}$$

where we denote (cf. also (1.1) and (1.2))

$$\mathbf{r}_R \equiv \mathbf{d}_R - \mathrm{Div}\,\mathbf{T}_R = \varrho_R\ddot{\chi} - \varrho_R\mathbf{b} - \mathrm{Div}\,\mathbf{T}_R , \quad \mathbf{s}_R \equiv \mathbf{t}_R + \mathbf{T}_R\mathbf{n}_R = \mathbf{T}_R\mathbf{n}_R - \mathbf{p}_R \tag{3.7}$$

and where $\mathbf{n}_R \equiv (n_{R\alpha})$ is a unit vector normal to ∂B_R ; relation (3.6) has to be satisfied for any $\delta\chi$ (cf. (3.4), (3.5)). We deal now with the mechanics of elastic media with constraints imposed on the deformation function only.

If Eqs. (3.1) have the form $\nabla\chi - \mathbf{F} = \mathbf{0}$, then from (3.3), using the divergence theorem and via du Bois-Reymonde lemma, we obtain $\mathbf{d}_R = \mathrm{Div}\,\mathbf{T}_R$, $\mathbf{t}_R = -\mathbf{T}_R\mathbf{n}_R$.

(*) The solution of this equation is not unique.

Hence we see that Eqs. (1.1), (1.2) and (2.1) take the well known form of the basic equations of the classical non-linear elasticity.

4. Material Continua with Kinetic Restrictions. Ideal Restrictions for the Kinetic Fields. Passage to the Classical Non-linear Elasticity.

In Sec. 3 the unknown internal forces $\mathbf{d}_R$, $\mathbf{t}_R$, $\mathbf{T}_R$ were interpreted as the reaction forces which maintain the postulated kinematic restrictions $h_\nu(\cdot)=0$ and $R_\varrho(\cdot)=0$.

Now we are to assume that there are no kinematic restrictions and that the internal forces $\mathbf{d}_R$, $\mathbf{t}_R$, $\mathbf{T}_R$ can be interpreted exclusively as the forces of the interparticle interactions in the material continuum. Such system of forces has to be self-equilibrated due to the known action and reaction principle. The necessary (global) conditions which are implied by the latter requirement have the form

$$\oint_{\partial B_R} \mathbf{t}_R \, ds_R + \int_{B_R} \mathbf{d}_R \, dv_R = \mathbf{0} \quad ; \quad \oint_{\partial B_R} \mathbf{t}_R \times \chi \, ds_R + \int_{B_R} \mathbf{d}_R \times \chi \, dv_R = \mathbf{0} \tag{4.1}$$

Moreover, if we assume that the local form of the principle of action and reaction, which relates the fields $\mathbf{t}_R$, $\mathbf{T}_R$ on ∂B_R and the fields $\mathbf{d}_R$, $\mathbf{T}_R$ in B_R, is included into the equations of restrictions in B, these will be given by

$$h^\mu(\mathbf{X}, t\,; \mathbf{d}_R, \mathbf{T}_R, \nabla\mathbf{T}_R, \nabla\mathbf{d}_R, \pi, \nabla\pi) = 0\,, \quad \mu = 1, \ldots, \bar{s}\,, \tag{4.2}$$

and on ∂B_R they are assumed to take the form

$$R^\sigma(\mathbf{X}, t, \mathbf{t}_R, \mathbf{T}_R, \pi) = 0\,, \quad \sigma = 1, \ldots, \bar{\bar{s}}\,, \tag{4.3}$$

where h^μ R^σ are known differentiable functions of all arguments and $\pi = (\pi^c(\mathbf{X},t))$, $c = 1, \ldots, m$ is an unknown vector function defined on $\bar{B}_R \times R$. The meaning of the vector π is quite analogous to that of the vector ψ in Eqs. (3.1) and (3.2). Moreover, on the relations (4.1), (4.3) we impose the same analytical conditions as on the relations (3.1), (3.2). In order to determine the relation between the form of kinetic restrictions (4.2)-(4.3) and the kinematic behavior of the continuum, we postulate that the kinetic restrictions (4.2)-(4.3) are ideal ; it means that the relation

$$\oint_{\partial B_R} \chi\cdot\delta \mathbf{t}_R \, ds_R + \int_{B_R} (\chi\cdot\delta \mathbf{d}_R + \mathbf{F}\cdot\delta \mathbf{T}_R) \, dv_R = 0 \tag{4.4}$$

holds for any $\delta \mathbf{t}_R = (\delta t^k_R)$, $\delta \mathbf{d}_R = (\delta d^k_R)$, $\delta \mathbf{T}_R = (\delta T^{k\alpha}_R)$,(*) such that $\delta \mathbf{t}_R, \delta \mathbf{d}_R, \delta \mathbf{T}_R$, $\delta \pi$) is a solution of the following system of the partial differential equations in B_R

$$\frac{\partial h^\mu}{\partial d^k_R}\delta d^k_R + \frac{\partial h^\mu}{\partial d^k_{R,\alpha}}\delta d^k_{R,\alpha} + \frac{\partial h^\mu}{\partial T^{k\alpha}_R}\delta T^{k\alpha}_R + \frac{\partial h^\mu}{\partial T^{k\alpha}_{R,\beta}}\delta T^{k\alpha}_{R,\beta} + \frac{\partial h^\mu}{\partial \pi^c}\delta\pi^c + \\ + \frac{\partial h^\mu}{\partial \pi^c_{,\alpha}}\delta\pi^c_{,\alpha} = 0 , \tag{4.5}$$

and the boundary conditions on ∂B_R

$$\frac{\partial R^\varrho}{\partial t^k_R}\delta t^k_R + \frac{\partial R^\varrho}{\partial T^{k\alpha}_R}\delta T^{k\alpha}_R + \frac{\partial R^\varrho}{\partial \pi^c}\delta\pi^c = 0 . \tag{4.6}$$

Eqs. (4.5) and (4.6) are defined for any $(\mathbf{t}_R, \mathbf{d}_R, \mathbf{T}_R, \pi)$ satisfying (4.2)-(4.3). Mind, that the form (4.2), (4.3) of kinetic restrictions has to ensure via (4.4) the existence of the fields χ , $\mathbf{F}$ satisfying the conditions given in Secs. 1, 2.

If among the equations (4.2), (4.3) there are the equations $\mathbf{d}_R - \mathrm{Div}\,\mathbf{T}_R = \mathbf{0}$, $\mathbf{t}_R + \mathbf{T}_R\mathbf{n}_R = \mathbf{0}$, respectively ($\mathbf{n}_R$ being the unit vector normal to ∂B_R), then after applying the divergence theorem to (4.4) we obtain

$$\int_{B_R} \mathbf{J}\cdot\delta \mathbf{T}_R \, dv_R = 0 , \tag{4.7}$$

where we have denoted

$$\mathbf{J} = (J_{k\alpha}) = \nabla\chi - \mathbf{F} . \tag{4.8}$$

(*) For the physical interpretation of Eq. (4.4) cf. [5] .

The relation (4.7) has to be sastified for any $\delta \mathbf{T}_R$. If $\mathbf{J} \neq \mathbf{O}$ we deal with the mechanics of slender bodies based on the concept of kinetic restrictions.

If equations (4.2) and (4.3) reduce to $\mathbf{d}_R - \operatorname{Div} \mathbf{T}_R = \mathbf{O}$ and $\mathbf{t}_R + \mathbf{T}_R \mathbf{n}_R = \mathbf{O}$, respectively, then from (4.4) we obtain $\mathbf{F} = \nabla \chi$; it is easy to see that this special case of kinetic restrictions leads directly to the classical non-linear theory of elasticity.

Let us now assume that we deal with the kinetic restrictions related to the present configuration, given by

(4.9) $$h^{\mu}(\mathbf{X}, t, \mathbf{d}, \overline{\mathbf{T}}, \nabla \mathbf{d}, \nabla \overline{\mathbf{T}}, \pi, \nabla \pi) = 0, \quad \mu = 1,\dots,\bar{\nu}, \quad \mathbf{X} \in \partial B_R ,$$

and by

(4.10) $$R^{\varrho}(\mathbf{X}, t, \mathbf{t}, \overline{\mathbf{T}}, \pi) = 0, \quad \varrho = 1,\dots,\bar{s}, \quad \mathbf{X} \in \partial B_R .$$

From (4.4) we obtain

(4.11) $$\oint_{\partial B_t} \chi \cdot \delta \mathbf{t} \, ds + \int_{B_t} (\chi \cdot \delta \mathbf{d} + \mathbf{C} \cdot \delta \overline{\mathbf{T}}) \, dv = 0$$

for any $\delta \mathbf{t}$, $\delta \mathbf{d}$, $\delta \overline{\mathbf{T}}$, such that $(\delta \mathbf{t}, \delta \mathbf{d}, \delta \overline{\mathbf{T}}, \delta \pi)$ is a solution of the following system of equations in B_R

$$\frac{\partial h^{\mu}}{\partial d^k} \delta d^k + \frac{\partial h^{\mu}}{\partial d^k_{,\alpha}} \delta d^k_{,\alpha} + \frac{\partial h^{\mu}}{\partial \overline{T}^{\alpha\beta}} \delta \overline{T}^{\alpha\beta} + \frac{\partial h^{\mu}}{\partial \overline{T}^{\alpha\beta}_{,\gamma}} \delta \overline{T}^{\alpha\beta}_{,\gamma} + \frac{\partial h^{\mu}}{\delta \pi^c} \delta \pi + \frac{\partial h^{\mu}}{\delta \pi^c_{;\alpha}} \delta \pi^c_{;\alpha} = 0$$

(4.12)

and the conditions on ∂B_R given by

(4.13) $$\frac{\partial R^{\varrho}}{\partial t^k} \delta t^k + \frac{\partial R^{\varrho}}{\partial T^{\alpha\beta}} \delta \overline{T}^{\alpha\beta} + \frac{\partial R^{\varrho}}{\partial \pi^c} \delta \pi^c = 0 .$$

Equations (4.12) and (4.13) are defined for any $(\mathbf{t}, \mathbf{d}, \overline{\mathbf{T}}, \pi)$ satisfying (4.9), (4.10). At the same time Eqs. (4.9) and (4.10) can not be quite arbitrary because not all fields $\mathbf{t}$, $\mathbf{d}$, $\overline{\mathbf{T}}$ are admissible when we deal with the ideal kinetic restrictions defined by (4.11) (i.e. Eqs. (4.9)-(4.13) have to ensure the existence of the fields χ , $\mathbf{F}$ introduced in Secs. 1, 2).

5. Some General Theorems. Principle of Virtual Work and Principle of Complementary Virtual Work.

To construct the analytical mechanics of elastic media we have to postulate equations of dynamics (1.1) and (1.2), to characterize the external loads by means of (1.3), to define the elastic material by specifying the strain energy function $\sigma = \hat{\sigma}(\mathbf{X},\mathbf{C})$ in (2.3) and then to introduce certain ideal restrictions imposed either on the kinematic fields (using Eqs. (3.1)-(3.3)) or on the kinetic fields (by taking into account (4.2), (4.3), (4.4)).

From the basic axioms listed here we shall obtain in Secs. 6 and 7 the field equations and the kinetic boundary conditions of the analytical mechanics of elastic media. In this Section we are to present some general theorems.

We shall start with the theorems concerning the elastic continuum based on the ideal kinematic restrictions. If Eqs. (3.1), (3.2) are invariant under any rigid translation of the physical space, then the resultant of all internal interactions and surface tractions is equal to zero :

$$\oint_{\partial B_R} \mathbf{t}_R \, d\sigma_R + \int_{B_R} \mathbf{d}_R \, dv_R = 0 . \tag{5.1}$$

To prove this statement it is sufficient to show that Eq. (3.3) holds for any constant vector $\delta\chi = \mathbf{c}$, representing the rigid translation of the physical space, and that the assumed invariance condition is satisfied only if h_ν and R_ϱ do not depend explicitly on χ .

In the same way we can prove that if the form of kinematic restrictions is invariant under any rigid rotation of the physical space, then the resultant of all internal interactions and surface tractions is equal to zero :

$$\oint_{\partial B_R} \mathbf{t}_R \times \chi \, d\sigma_R + \int_{B_R} \mathbf{d}_R \times \chi \, dv_R = 0 . \tag{5.2}$$

To prove this theorem we have also to take into account that $T_R^{[k\alpha} F_\alpha^{l]} = 0$, which follows from (2.3).

If the kinematic restrictions are scleronomic (the functions h_ν and R_ϱ do not depend on t explicity), then the work done by all reaction forces $\mathbf{t}_R$, $\mathbf{d}_R$, $\mathbf{T}_R$ is zero:

$$\oint_{\partial B_R} \mathbf{t}_R \cdot \dot{\chi}\, d\sigma_R + \int_{B_R} (\mathbf{d}_R \cdot \dot{\chi} + \mathbf{T}_R \cdot \dot{\mathbf{F}})\, dv_R = 0 . \tag{5.3}$$

Under the same assumptions as concerning (5.1)-(5.3) and using (1.1), (1.2) and (2.1) we shall prove that the following laws of conservation of momentum, moment of momentum and energy hold :

$$\frac{d}{dt} \int_{B_R} \varrho_R \dot{\chi}\, dv_R = \oint_{\partial B_R} \mathbf{p}_R\, d\sigma_R + \int_{B_R} \varrho_R \mathbf{b}\, dv_R , \tag{5.4}$$

$$\frac{d}{dt} \int_{B_R} \varrho_R \dot{\chi} \times \chi\, dv_R = \oint_{\partial B_R} \mathbf{p}_R \times \chi\, d\sigma_R + \int_{B_R} \varrho_R \mathbf{b} \times \chi\, dv_R , \tag{5.5}$$

$$\frac{d}{dt} \int_{B_R} \left(\frac{1}{2} \varrho_R |\dot{\chi}|^2 + \varrho_R \sigma\right) dv_R = \oint_{\partial B_R} \mathbf{p}_R \cdot \dot{\chi}\, d\sigma_R + \int_{B_R} \varrho_R \mathbf{b} \cdot \dot{\chi}\, dv_R . \tag{5.6}$$

Relations of the form (5.4)-(5.6) are valid only for the whole dynamic system ; we can see that the reaction forces $\mathbf{t}_R$, $\mathbf{d}_R$, $\mathbf{T}_R$ due to the kinematic restrictions do not interfere in the conservation laws in their global form. The laws of conservation for an arbitrary part of the continuum can be obtained directly from (1.1).

Substituting right-hand sides of (1.1) and (1.2) into (3.3) we conclude that the restrictions imposed on the kinematic fields χ , $\mathbf{F}$ are ideal if and only if the following principle of virtual work

$$\oint_{\partial B_R} \mathbf{p}_R \cdot \delta\chi\, d\sigma_R + \int_{B_R} \varrho_R (\mathbf{b} - \ddot{\chi}) \cdot \delta\chi\, dv_R = \int_{B_R} \mathbf{T}_R \cdot \delta\mathbf{F}\, dv_R \tag{5.7}$$

holds for any $\delta\chi$, $\delta\mathbf{F}$. The inverse statement also holds.

Let us consider the mechanics of continuum with kinetic restrictions. We can formulate the principle dual to the principle of virtual work (5.7), taking into account the condition (4.4) and the dynamic equations (1.1) and (1.2). This principle will be called the principle of complementary virtual work ; we assert that the following relation

$$\oint_{\partial B_R} \boldsymbol{\chi} \cdot \delta \mathbf{p}_R \, d\sigma_R + \int_{B_R} \varrho_R \boldsymbol{\chi} \cdot \delta \mathbf{f} \, dv_R = \int_{B_R} \mathbf{F} \cdot \delta \mathbf{T}_R \, dv_R \,, \quad \mathbf{f} = \mathbf{b} - \ddot{\boldsymbol{\chi}} \tag{5.8}$$

holds for any $\delta \mathbf{p}_R$, $\delta \mathbf{f}$, $\delta \mathbf{T}_R$, such that $\delta \mathbf{p}_R + \delta \mathbf{t}_R = \mathbf{0}$, $\varrho_R \delta \mathbf{f} + \delta \mathbf{d}_R = \mathbf{0}$ and $\delta \mathbf{t}_R$ $\delta \mathbf{d}_R$, $\delta \mathbf{T}_R$ satisfy the system (4.4), (4.5). Assuming that there exists the function of complementary energy $\gamma = \gamma(\mathbf{X}, \bar{\mathbf{T}})$, the stress relations (2.3) can be written in an inverse form $\mathbf{C} = \partial\gamma / \partial \bar{\mathbf{T}}$ and we obtain an alternative form of the principle of complementary virtual work

$$\oint_{\partial B_t} \boldsymbol{\chi} \cdot \delta \mathbf{p} \, d\sigma + \int_{B_t} \varrho \boldsymbol{\chi} \cdot d\mathbf{f} \, dv = \int_{B_t} d\gamma \, dv \,, \tag{5.9}$$

where $\delta \mathbf{p} + \delta \mathbf{t} = \mathbf{0}$, $\varrho \delta \mathbf{f} + \delta \mathbf{d} = \mathbf{0}$ and where $\delta\gamma$ is a variation of the complementary energy function ; at the same time Eqs. (4.5), (4.6) have to be taken into account.

6. Lagrange's Equations of the First Kind.

Using the known Lagrange's multipliers approach we can obtain from (3.3) and (1.1), (1.2) the equivalent system of the field equations and kinetic boundary conditions. To this aid we denote by $\lambda^\nu = \lambda^\nu(\mathbf{X}, t)$, $\mathbf{X} \in B_R$, $t \in R$ the internal Lagrange's multipliers corresponding to Eqs. (3.1) and by $\mu^\varrho = \mu^\varrho(\mathbf{X}, t)$, $\mathbf{X} \in \partial B_R$, $t \in R$ the boundary Lagrange's multipliers connected with Eqs. (3.2). After applying to (5.3) the Lagrange's multipliers approach, we obtain the following system of field equations in $B_R \times R$

$$\begin{gathered} \operatorname{Div}\left(\lambda^\nu \frac{\partial h_\nu}{\partial \nabla \boldsymbol{\chi}}\right) - \lambda^\nu \frac{\partial h_\nu}{\partial \boldsymbol{\chi}} + \varrho_R \mathbf{b} = \varrho_R \ddot{\boldsymbol{\chi}} \,, \quad \mathbf{T}_R + \lambda^\nu \frac{\partial h_\nu}{\partial \mathbf{F}} = \mathbf{0} \,, \\ \operatorname{Div}\left(\lambda^\nu \frac{\partial h_\nu}{\partial \nabla \boldsymbol{\psi}}\right) - \lambda^\nu \frac{\partial h_\nu}{\partial \boldsymbol{\psi}} = 0 \,, \end{gathered} \tag{6.1}$$

and the kinetic boundary conditions on $\partial B_R \times R$

$$\lambda^{\nu}\frac{\partial h_{\nu}}{\partial \nabla\chi}\mathbf{n}_R = \mathbf{p}_R + \mu^{\varrho}\frac{\partial R_{\varrho}}{\partial \chi}\,, \quad \lambda^{\nu}\frac{\partial h_{\nu}}{\partial \nabla\psi}\mathbf{n}_R = \mu^{\varrho}\frac{\partial R_{\varrho}}{\partial \psi}\,, \tag{6.2}$$

to obtain the foregoing relations we eliminate $\mathbf{d}_R$, $\mathbf{t}_R$ using (1.1) and (1.2) and at the same time we assume that the conditions of regularity needed to apply the theorem of the Lagrange's multipliers are satisfied.

If the equations of kinematic restrictions have the form

$$\begin{gathered} \nabla\chi - \mathbf{F} = \mathbf{0} \\ \bar{h}_{\nu}(\mathbf{X}, t, \chi, \nabla\chi, \psi, \nabla\psi) = 0\,, \end{gathered} \tag{6.3}$$

then denoting by $\boldsymbol{\Lambda} = (\Lambda^{k\alpha})$ the Lagrange's multipliers corresponding to the Eqs. $(6.3)_1$ and by $\bar{\lambda}^{\nu}$ the Lagrange's multipliers corresponding to Eqs. $(6.3)_2$, we obtain from $(6.1)_2$ that $\boldsymbol{\Lambda} = \mathbf{T}_R$ and that the equations (6.1) reduce to

$$\begin{gathered} \operatorname{Div}\left(\mathbf{T}_R + \bar{\lambda}^{\nu}\frac{\partial h_{\nu}}{\partial \nabla\chi}\right) - \bar{\lambda}^{\nu}\frac{\partial \bar{h}_{\nu}}{\partial \chi} + \varrho_R \mathbf{b} = \varrho_R \ddot{\chi}\,, \\ \operatorname{Div}\left(\bar{\lambda}^{\nu}\frac{\partial \bar{h}_{\nu}}{\partial \nabla\psi}\right) - \bar{\lambda}^{\nu}\frac{\partial \bar{h}_{\nu}}{\partial \psi} = 0\,, \end{gathered} \tag{6.4}$$

while the kinetic boundary conditions (6.2) will be given by

$$\left(\mathbf{T}_R + \bar{\lambda}^{\nu}\frac{\partial \bar{h}_{\nu}}{\partial \nabla\chi}\right)\mathbf{n}_R = \mathbf{p}_R + \mu^{\varrho}\frac{\partial R_{\varrho}}{\partial \chi}\,; \quad \bar{\lambda}^{\nu}\frac{\partial \bar{h}_{\nu}}{\partial \nabla\psi}\mathbf{n}_R = \mu^{\varrho}\frac{\partial R_{\varrho}}{\partial \psi}\,. \tag{6.5}$$

The equations (6.1) will be called the Lagrange's equations of motion of the first kind ; at the same time we shall refer to Eqs. (6.2) as the kinetic boundary conditions of the first kind. The equations stated above, the restrictions imposed on the kinematic fields χ, $\mathbf{F}$, and Eqs. (1.3), (2.1), describe the mechanics of elastic "structured" continuum. In this approach the dynamic equations (1.1) and (1.2) can be used to determine the fields $\mathbf{d}_R$, $\mathbf{t}_R$ of reaction forces.

Let us now consider the case in which the restrictions (4.9), (4.10) imposed on the kinetic fields ; however, we confine ourselves to the special case of these restrictions given in

$$\mathbf{d}-\hat{\mathbf{d}}\left(\mathbf{X},t,\nabla\bar{\mathbf{T}}\right)=0$$

$$\bar{h}^{\mu}\left(\mathbf{X},t,\bar{\mathbf{T}},\nabla\bar{\mathbf{T}},\pi,\nabla\pi\right)=0\ ;\ \mathbf{X}\in B_R\ , \tag{6.6}$$

$$\mathbf{t}-\hat{\mathbf{t}}\left(\mathbf{X},t,\bar{\mathbf{T}}\right)=0\ ;\ \mathbf{X}\in\partial B_R\ ;\ \nabla\bar{\mathbf{T}}=\left(\bar{T}^{\alpha\beta}\big|_{\gamma}\right),$$

where $\hat{\mathbf{d}}$ and $\hat{\mathbf{t}}$ are known differentiable functions. Substituting $(6.6)_{1,3}$ into (4.11) we obtain

$$\oint_{\partial B_t}\left(\chi\cdot\frac{\partial\hat{\mathbf{t}}}{\partial\bar{T}^{\alpha\beta}}+\chi\cdot\frac{\partial\mathbf{d}}{\partial\bar{T}^{\alpha\beta}|_{\gamma}}n_{\gamma}\right)\delta\bar{T}^{\alpha\beta}d\sigma+\int_{B_t}\left[C_{\alpha\beta}-\left(\chi\cdot\frac{\partial\hat{\mathbf{d}}}{\partial\bar{T}^{\alpha\beta}|_{\gamma}}\right)\bigg|_{\gamma}\right]\delta\bar{T}^{\alpha\beta}dv=0. \tag{6.7}$$

Now we apply the Lagrange's multipliers approach to (6.7) and $(6.6)_2$. Denoting by $\bar{\lambda}_{\mu}=\bar{\lambda}_{\mu}(\mathbf{X},t)$ the Lagrange's multipliers corresponding to Eqs. $(6.6)_2$, we obtain the fields equations in $B_R\times R$

$$C_{\alpha\beta}-\left(\chi\cdot\frac{\partial\hat{\mathbf{d}}}{\partial\bar{T}^{\alpha\beta}|_{\gamma}}+\bar{\lambda}_{\mu}\frac{\partial\bar{h}^{\mu}}{\partial\bar{T}^{\alpha\beta}|_{\gamma}}\right)\bigg|_{\gamma}+\bar{\lambda}_{\mu}\frac{\partial\bar{h}^{\mu}}{\partial\bar{T}^{\alpha\beta}}=0\quad,\quad\left(\bar{\lambda}_{\mu}\frac{\partial\bar{h}^{\mu}}{\partial\pi^{c}_{;\alpha}}\right)_{,\alpha}-\bar{\lambda}_{\mu}\frac{\partial h^{\mu}}{\partial\pi^{c}}=0\ , \tag{6.8}$$

and the boundary conditions on $\partial B_R\times R$

$$\chi\cdot\frac{\partial\hat{\mathbf{t}}}{\partial\bar{T}^{\alpha\beta}}+\chi\cdot\frac{\partial\hat{\mathbf{d}}}{\partial\bar{T}^{\alpha\beta}|_{\gamma}}n_{\gamma}+\bar{\lambda}_{\mu}\frac{\partial\bar{h}^{\mu}}{\partial\bar{T}^{\alpha\beta}|_{\gamma}}n_{\gamma}=0\ ,\quad\bar{\lambda}_{\mu}\frac{\partial\bar{h}^{\mu}}{\partial\pi^{c}_{;\alpha}}n_{\alpha}=0\ . \tag{6.9}$$

The equations (6.8) are called the Lagrange's compatibility field conditions of the first kind and Eqs. (6.9) are said to be compatibility boundary conditions of the first kind.

The compatibility equations (6.8), (6.9), the kinetic restrictions (6.6), the equations of dynamics (1.1), (1.2), which related to the present configuration have the form

$$\varrho\ddot{\chi}=\mathbf{d}+\varrho\mathbf{b}\ ;\quad\mathbf{X}\in B_R\ , \tag{6.10}$$

and

$$\mathbf{t} + \mathbf{p} = \mathbf{0} \; ; \quad \mathbf{X} \in \partial B_R \; , \tag{6.11}$$

respectively, with the equations which define the external loads

$$\mathbf{p} = \hat{\pi}[\chi] \; , \quad \mathbf{b} = \hat{\beta}[\chi] \tag{6.12}$$

and with the stress relation (2.3), describe the mechanics of the material continuum with kinetic restrictions of the form (6.6).

If the number of equations of constraints $h_\nu = 0$ or $\bar{h}^\mu = 0$ is large, then the Lagrange's equations of the first kind depend on the large number of unknown multipliers λ^ν or $\bar{\lambda}_\mu$ and the solution of the particular boundary-value problems is very involved. In actual practice these equations are applied only if the number of equations $h_\nu = 0$ or $h^\mu = 0$ is small ; for example the kinematic restrictions are used to describe the mechanics of incompressible materials or materials reinforced with flexible and non-extensible families of cords. More complicated problems are investigated by means of the concept of generalized coordinates and by that of the generalized forces ; the suitable equations are said to be the Lagrange's equations of the second kind ; this approach will be studied in the next Section.

7. Lagrange's Equations of the Second Kind.

When we construct theories of shells, plates, membrans or rods, or when we deal with the finite element approach, as well as with the variational approaches, the kinematic-kinetic restrictions will be postulated in the special form which will be called explicit. It is the form in which unknown kinematic χ , $\mathbf{F}$ or kinetic $\mathbf{t}_R$, $\mathbf{d}_R$, $\mathbf{T}_R$ fields are expressed in terms of a certain unknown vector function, components of which for the kinematical restrictions are called the generalized coordinates (or the generalized deformations) and for the kinetic restrictions they are said to be the generalized forces. If such situation occurs then the basic equations which can be obtained from the suitable ideality condition do not depend on the Lagrange's multipliers.

Let us assume that from Eqs. (3.1) we can obtain χ and $\mathbf{C}=\mathbf{F}^T\mathbf{F}$ in terms of ψ and let the derivatives of ψ with respect to one fixed material coordinate be equal to zero. It is a case in which we deal with the following special form of Eqs. (3.3) :

$$C_{\alpha\beta}=\Phi_{\alpha\beta}\big(\mathbf{X},t\,;\,\psi(\mathbf{Z},t),\,\nabla\psi(\mathbf{Z},t)\big),$$
$$\chi_k=\varphi_k\big(\mathbf{X},t\,;\psi(\mathbf{Z},t)\big);\quad \mathbf{Z}=\pi(\mathbf{X});\quad \mathbf{X}\in B_R,\ t\in R, \tag{7.1}$$

where $\Phi_{\alpha\beta}$, φ_k are known differentiable functions (we have assumed that φ is independent of $\nabla\psi$) and $\pi:\mathbf{B}_R\rightarrow\Pi$ is a given differentiable projection. The components $\psi_a(\mathbf{Z},t)$ of the vector ψ are called the generalized coordinates. We are to obtain the system of equations for the generalized coordinates $\psi_a(\mathbf{Z},t)$, $\mathbf{Z}\in\bar{\Pi}$, $t\in R$, $a=1,..,n$; at the same time we assume that the generalized coordinates have to satisfy certain conditions given by

$$\alpha_\varrho(\mathbf{Z},t,\psi)=0\;;\quad \mathbf{Z}\in\partial\Pi,\ t\in R\;;\ \varrho=1,\dots,s, \tag{7.2}$$

where α_ϱ are known differentiable functions. To make our analysis more simple we shall assume that Π is a region on the plane R^2 parametrized with Cartesian orthogonal coordinates $\mathbf{Z}=(Z^K)$, $K=1,2$, and we denote $E\equiv\pi^{-1}(\mathbf{Z})$, $\mathbf{Z}\in\Pi$.

Because of $dv=\mathfrak{J}\,dE\,d\Pi$ and after substituting the right-hand sides of the following equalities into (3.8)

$$\delta\mathbf{C}=\frac{\partial\Phi}{\partial\psi_a}\delta\psi_a+\frac{\partial\Phi}{\partial\psi_{a,K}}\delta\psi_{a,K},\quad \delta\chi=\frac{\partial\varphi}{\partial\psi_a}\delta\psi_a\;;\ a=1,\dots,n\;;\ K=1,2, \tag{7.3}$$

by virtue of the divergence theorem we obtain

$$\int_\Pi\oint_{\partial E}\Big(\mathbf{t}_R\cdot\frac{\partial\varphi}{\partial\psi_a}+\frac{1}{2}\mathfrak{J}\bar{\mathbf{T}}\cdot\frac{\partial\Phi}{\partial\psi_{a,\alpha}}n_\alpha\Big)d(\partial E)\,\delta\psi_a\,d\Pi\,+$$
$$+\int_\Pi\int_E\Big[\mathbf{d}_R\cdot\frac{\partial\varphi}{\partial\psi_a}+\frac{1}{2}\mathfrak{J}\bar{\mathbf{T}}\cdot\frac{\partial\Phi}{\partial\psi_a}-\frac{1}{2}\Big(\mathfrak{J}\bar{\mathbf{T}}\cdot\frac{\partial\Phi}{\partial\psi_{a,\alpha}}\Big)_{,\alpha}\Big]\delta\psi_a\,dE\,d\Pi\,+$$
$$+\oint_{\partial\Pi}\int_E\Big(\mathbf{t}_R\cdot\frac{\partial\varphi}{\partial\psi_a}+\frac{1}{2}\bar{\mathbf{T}}\cdot\frac{\partial\Phi}{\partial\psi_{a,\alpha}}n_\alpha\Big)dE\,\delta\psi_a\,d(\partial\Pi)=0\,. \tag{7.4}$$

We can apply to (7.4) the du Bois-Reymonde lemma, obtaining the system of the field equations and the boundary conditions on $\partial\Pi$. Using (1.1) and (1.2) this system can be written down in the form

$$\frac{1}{2}\Big(\int_E \mathfrak{J}\bar{\mathbf{T}}\cdot\frac{\partial\Phi}{\partial\psi_{a,K}}dE\Big)_{,K}-\frac{1}{2}\int_E \mathfrak{J}\bar{\mathbf{T}}\cdot\frac{\partial\Phi}{\partial\psi_a}dE+f^a=\frac{d}{dt}\frac{\partial\varkappa}{\partial\dot{\psi}_a}-\frac{\partial\varkappa}{\partial\psi_a}\,, \tag{7.5}$$

and

$$\frac{1}{2}\int_E \mathfrak{J}\bar{\mathbf{T}}\cdot\frac{\partial\Phi}{\partial\psi_{a,K}}dE\,n_K=p^a+\mu^\varrho\frac{\partial\alpha_\varrho}{\partial\psi_a}\,, \tag{7.6}$$

where $\mu^\varrho=\mu^\varrho(\mathbf{Z},t),\mathbf{Z}\in\partial\Pi$, $t\in R$, are the Lagrange's multipliers for Eqs. (7.2) on $\partial\Pi$, and where we have denoted

$$f^a=\int_{\partial E}\pi_R[\varphi]\cdot\frac{\partial\varphi}{\partial\psi_a}d(\partial E)+\int_E\varrho\beta[\varphi]\cdot\frac{\partial\varphi}{\partial\psi_a}dE\,,$$

$$\varkappa=\frac{1}{2}\int_E\varrho_R|\dot{\varphi}|^2dE\,;\qquad E=\big(a(z),b(z)\big)\subset R\,, \tag{7.7}$$

$$p^a=\int_E\pi_R[\varphi]\cdot\frac{\partial\varphi}{\partial\psi_a}dE\,,\qquad \oint_{\partial E}(\ )d(\partial E)=(\ )\big|_{a(z)}+(\ )\big|_{b(z)}\,.$$

Now let us substitute to (7.5), (7.6) the right-hand side of the stress relation (2.3). After defining the function

$$\varepsilon=\varepsilon(\mathbf{Z},t;\psi,\nabla\psi)=\int_E\varrho_R\hat{\sigma}(\mathbf{X},\Phi)dE \tag{7.8}$$

we obtain finally the following field equations for the generalized coordinates

$$\Big(\frac{\partial\varepsilon}{\partial\psi_{a,K}}\Big)_{,K}-\frac{\partial\varepsilon}{\partial\psi_a}+f^a=\frac{d}{dt}\frac{\partial\varkappa}{\partial\dot{\psi}_a}-\frac{\partial\varkappa}{\partial\psi_a}\,,\qquad \mathbf{Z}\in\Pi\,,\ a=1,\dots,n \tag{7.9}$$

and the boundary conditions

$$\frac{\partial\varepsilon}{\partial\psi_{a,K}}n_K=p^a+\mu^\varrho\frac{\partial\alpha_\varrho}{\partial\psi_a}\,;\qquad \mathbf{Z}\in\partial\Pi\,,\quad a=1,\dots,n\,. \tag{7.10}$$

Equations (7.9), which are said to be Lagrange's equations of the second kind for ψ_a, with the boundary conditions (7.10) (the kinetic boundary conditions of the second kind) and (7.2), represent the boundary-value problem for the generalized coordinates.

In the same way we can obtain the Lagrange's equations of the second kind if Π is a straight line segment and E is a region on the plane. Using the same approach we can also arrive at the field equations if the generalized coordinates are only time dependent functions $\psi = (\psi_a(t))$; in this case we obtain the field equations in the form of Eqs. (7.9) but without the first term ; at the same time in Eqs. (7.7), (7.8) the integrals have to be taken over B_R .

The equations (7.9), (7.10) are also valid if the Eqs. $(7.1)_1$ are postulated not for each pair $(\alpha,\beta)=(\beta,\alpha)$ of subscripts but only for some pairs. Assuming $(7.1)_1$ in the form $C_{KL} = \Phi_{KL}(\cdot)$, $C_{33} = \Phi_{33}(\cdot)$ we obtain from (5.1) that $\bar{T}^{K3} = \bar{T}^{3K} = 0$. Using the stress relation (2.3) for $\bar{T}^{K3}$ we express the unknowns $C_{K3}(X,t)$ in terms of C_{KL} , C_{33} ; this enables us to determine next the function ε using (7.8).

Now we are to give another example of the Lagrange's equations of the second kind for the generalized coordinates. To this aim we postulate the restrictions in the form

$$C_{\alpha\beta} = \Phi_{\alpha\beta}\left(X, t ; \nabla\chi\right), \quad X \in B_R . \tag{7.11}$$

As the generalized coordinates ψ_a we can take here the components $\chi_k(X,t)$ of the vector χ (of the deformation function). Substituting

$$\delta C = \frac{\partial\Phi}{\partial\chi_{k,\alpha}}\, \partial\chi_{k,\alpha} \tag{7.12}$$

into (3.6) we obtain

$$\oint_{\partial B_t}\left(t + \frac{1}{2}\,\frac{\partial\Phi}{\partial\chi_{k,\alpha}}\cdot\bar{T}n_\alpha\right)\delta\chi\, d\sigma + \int_{B_t}\left[d - \frac{1}{2}\left(\frac{\partial\Phi}{\partial\chi_{k,\gamma}}\cdot\bar{T}\right)\Big|_\gamma\right]\delta\chi\, dv = 0 \tag{7.13}$$

After applying du Bois-Reymonde lemma we arrive at the following Lagrange's equations of the second kind for $\chi_k(X,t)$:

$$\frac{1}{2}\left(\bar{T}^{\alpha\beta}\,\frac{\partial\Phi_{\alpha\beta}}{\partial\chi_{k,\gamma}}\right)\Big|_\gamma + \varrho\, b^k = \varrho\,\ddot{\chi}^k \quad ; \quad X \in B_R \tag{7.14}$$

and at the following kinetic boundary conditions

(7.15) $$\frac{1}{2}\,\frac{\partial \Phi_{\alpha\beta}}{\partial \chi_{k,\gamma}}\,\bar{T}^{\alpha\beta} n_{\gamma} = p^{k} \;; \qquad X \in \partial B_R \;.$$

To obtain (7.14), (7.15) we have used the dynamic relations (1.1), (1.2) in the form $p^k + t^k = 0$, $d^k + \varrho b^k = \varrho \ddot{\chi}^k$, related to the present configuration. Making use of (2.3) and defining the function

(7.16) $$\bar{\sigma} = \bar{\sigma}(X, t, \nabla\chi) \equiv \varrho\,\hat{\sigma}(X, \Phi)$$

we obtain finally the Lagrange's equations in the form

(7.17) $$\left(\frac{\partial \bar{\sigma}}{\partial \chi_{k,\alpha}}\right)\Big|_{\alpha} + \varrho b^{k} = \varrho \ddot{\chi}^{k}$$

and the kinetic boundary conditions given by

(7.18) $$\frac{\partial \bar{\sigma}}{\partial \chi_{k,\alpha}}\, n_{\alpha} = p^{k} \,.$$

If $\Phi = \nabla\chi^{T}\nabla\chi$, we arrive at the equations of the classical non-linear elasticity. If the relation (7.11) holds only for some pairs of indices $(\alpha,\beta) = (\beta,\gamma)$, then from (3.8) we obtain $\bar{T}^{\gamma\delta} = 0$ for all other pairs $(\gamma,\delta) = (\delta,\gamma)$ and the stress relations $\partial\hat{\sigma} / \partial C_{\gamma\delta} = 0$ can be used to determine the components $C_{\gamma\delta}$ which are not related to $\nabla\chi$ by means of the equations of constraints.

To give an example of the Lagrange's equations of the second kind for generalized forces, let us postulate the kinetic restrictions in the form

(7.19) $$\begin{aligned} d &= \delta(X, t\,;\, \bar{T}, \nabla\bar{T})\,; \qquad X \in B_R \\ t &= \tau(X, t\,;\, \bar{T})\,; \qquad X \in \partial B_R \end{aligned}$$

As generalized forces we can use here the components of the convective stress tensor. From (4.11), by virtue of du Bois-Reymonde lemma we obtain (cf. also Eqs. (6.6)-(6.9)) the Lagrange's equations of the second kind for the generalized stresses :

$$C_{\alpha\beta} - \left(\chi \cdot \frac{\partial \delta}{\partial \overline{T}^{\alpha\beta}|_\gamma}\right)\Big|_\gamma = 0\,, \quad X \in B_R\,, \tag{7.20}$$

and the corresponding kinetic boundary conditions

$$\chi \cdot \frac{\partial \pi}{\partial \overline{T}^{\alpha\beta}} + \chi \cdot \frac{\partial \delta}{\partial \overline{T}^{\alpha\beta}|_\gamma}\, n_\gamma = 0\,, \quad X \in \partial B_R \tag{7.21}$$

Equations (7.20) and (7.21) can be also called the Lagrange's compatibility relations of the second kind.

As a second example of the Lagrange's equations of the second kind for the generalized forces, let us consider the following form of kinetic restrictions

$$\begin{aligned} t^k &= -\chi^k_{;\alpha}\, \overline{T}^{\alpha\beta} n_\beta\;; \quad X \in \partial B_R\;; \\ d^k &= \chi^k_{;\alpha}\, \overline{T}^{\alpha\beta}|_\beta\,, \\ \overline{T}^{\alpha\beta} &= \varphi^{\alpha\beta}(X, t, \pi, \nabla\pi, \nabla\nabla\pi)\,; \quad X \in B_R \end{aligned} \tag{7.22}$$

where $\pi = (\pi^c(X,t))$, $c = 1,\ldots,m$, $(X,t) \in \overline{B}_R \times R$ is an arbitrary differentiable vector function. According to Eqs. (7.22) we can take the components of the vector π as the generalized forces. After calculating δt^k, δd^k from (7.22) and substituting the suitable expressions to (4.11) we arrive at the relation (4.16) with a denotation (4.17). Substituting to (4.16) the right-hand sides of

$$\delta \overline{T}^{\alpha\beta} = \frac{\partial \varphi^{\alpha\beta}}{\partial \pi^c}\,\delta\pi^c + \frac{\partial \varphi^{\alpha\beta}}{\partial \pi^c_{;\gamma}}\,\delta\pi^c_{;\gamma} + \frac{\partial \varphi^{\alpha\beta}}{\partial \pi^c_{;\gamma\delta}}\,\delta\pi^c_{;\gamma\delta} \tag{7.23}$$

and applying du Bois-Reymonde lemma to the obtained integral relation, we arrive at

$$D_{\alpha\beta}\frac{\partial \varphi^{\alpha\beta}}{\partial \pi^c} - \left[D_{\alpha\beta}\frac{\partial \varphi^{\alpha\beta}}{\partial \pi^c_{;\gamma}} - \left(D_{\alpha\beta}\frac{\partial \varphi^{\alpha\beta}}{\partial \pi^c_{;\gamma\delta}}\right)_{,\delta}\right]_{,\gamma} = 0\,; \quad \mathbf{D} = \nabla\chi^T\nabla\chi - \mathbf{C} \tag{7.24}$$

Eqs. (7.24) represent the Lagrange's equations of the second kind for the generalized forces $\pi^c(X,t)$. If the generalized forces and their derivatives at the boundary ∂B_R are known, then the corresponding boundary conditions for the generalized forces will be satisfied as identities. The field equations (7.24) are also called the compatibility conditions of deformations .

Now let us assume that the known function satisfies the relations

$$\varphi^{\alpha\beta}\big|_{\beta} \equiv 0 \; ; \quad \mathbf{X} \in B_R \, , \tag{7.25}$$

for any differentiable vector π of the generalized forces. From (7.25) and $(7.22)_2$, (1.1) follows that $\mathbf{b} = \ddot{\chi}$; when the continuum is in rest no loads can be taken into account. The generalized forces $\pi^c(\mathbf{X},t)$ can be then called the stress functions and the equation (7.24) takes the form

$$C_{\alpha\beta}\frac{\partial\varphi^{\alpha\beta}}{\partial\pi^c} - \left[C_{\alpha\beta}\frac{\partial\varphi^{\alpha\beta}}{\partial\pi^c_{,\gamma}} - \left(C_{\alpha\beta}\frac{\partial\varphi^{\alpha\beta}}{\partial\pi^c_{,\gamma\delta}}\right)\Big|_{\delta}\right]\Big|_{\gamma} = 0 \, , \tag{7.26}$$

For the elastic materials in which the complementary energy $\gamma = \gamma(\mathbf{X},\bar{\mathbf{T}})$ exists, by virtue of $C_{\alpha\beta} = \partial\gamma/\partial\bar{T}^{\alpha\beta}$, we obtain from Eqs. (7.26) the system of equations for the stress functions. If Eqs. $(7.22)_3$ reduce to the following ones

$$\bar{T}^{\alpha\beta} = \varphi^{\alpha\beta}(\mathbf{X},t,\pi(\mathbf{X},t)) \, , \tag{7.27}$$

then from (7.26) it follows that

$$\frac{\partial\bar{\gamma}}{\partial\pi^c} \equiv 0 \, , \quad c = 1,\ldots,m < 6 \, , \tag{7.28}$$

where we have denoted

$$\bar{\gamma} = \bar{\gamma}(\mathbf{X},t;\pi) \equiv \gamma(\mathbf{X},\boldsymbol{\varphi}(\mathbf{X},t;\pi)) \, . \tag{7.29}$$

If in (7.22) instead of $(7.22)_3$ we have

$$\bar{T}^{\alpha\beta} = \varphi^{\alpha\beta}(\mathbf{X},t;\pi(\mathbf{Z},t)) \tag{7.30}$$

where $X = (Z,Y) \in \Pi \times E = B_R$ and where the generalized forces $\pi^c(Z,T)$ are the stress functions, then we shall obtain

$$\frac{\partial}{\partial \pi^c} \int_E \bar{\gamma}\, dE = 0 . \tag{7.31}$$

In Eqs. (7.28) and (7.31) we can recognize the well known theorems concerning the complementary energy.

If the equations of constraints (7.22) are postulated only for some of the pairs $(\alpha,\beta) = (\beta,\gamma)$ of superscripts, then for any other pair $(\gamma,\delta) = (\delta,\gamma)$ from (4.16) we obtain $D_{\gamma\delta} = 0$ or $C_{\gamma\delta} = \chi^k_{,\gamma}\, \chi_{k,\delta}$. Thus the tensor $\mathbf{D} = (D_{\alpha\beta})$ can be called the incompatibility deformation tensor.

At the end of this contribution let us observe that there may exist many special cases of constraints in which we deal with the Lagrange's equations of the second kind with the Lagrange's multiplier. As an example let us take the constrains (7.1) and let us introduce the extra assumption that the generalized coordinates $\psi_a(\mathbf{Z},t)$ are not independent on each other but that they have to satisfy a system of equations in $\Pi \times R$

$$\gamma_\nu(\mathbf{Z},t;\ \psi(\mathbf{Z},t),\ \nabla\psi(\mathbf{Z},t)) = 0\ ,\quad \nu = 1,\ldots,r\ , \tag{7.32}$$

where γ_ν are known differentiable functions. Then using the Lagrange's multipliers method, instead of (7.4) and (7.5) we arrive at the following field equations

$$\left(\frac{1}{2}\int_E \mathfrak{J}\bar{\mathbf{T}}\cdot\frac{\partial\Phi}{\partial\psi_{a,K}}\,dE + \lambda^\nu\frac{\partial\gamma_\nu}{\partial\psi_{a,K}}\right)_{,K} - \frac{1}{2}\int_E \mathfrak{J}\bar{\mathbf{T}}\cdot\frac{\partial\Phi}{\partial\psi_a}\,dE - \lambda^\nu\frac{\partial\gamma_\nu}{\partial\psi_a} + f^a =$$

$$= \frac{d}{dt}\frac{\partial\varkappa}{\partial\dot{\psi}_a} - \frac{\partial\varkappa}{\partial\psi_a}\ ;\quad \lambda^\nu = \lambda^\nu(\mathbf{Z},t),\ \ \mathbf{Z}\in\Pi\ ,\ t\in R\ , \tag{7.33}$$

and at the subsequent kinetic boundary conditions

$$\left(\frac{1}{2}\int_E \mathfrak{J}\bar{\mathbf{T}}\cdot\frac{\partial\Phi}{\partial\psi_{a,K}}\,dE + \lambda^\nu\frac{\partial\gamma_\nu}{\partial\psi_{a,K}}\right)n_K = p^a + \mu^Q\frac{\partial a_Q}{\partial\psi_a}\ ;\ \ \mathbf{Z}\in\partial\Pi\ ,\ t\in R\ , \tag{7.34}$$

Taking into account the stress relation (2.3) we can obtain from (7.33) and (7.34) relations of the same form as that given by (8.8) and (8.9), respectively, but with the energy function ε defined not by Eqs. (8.7) but by the following equality

$$\varepsilon = \varepsilon(\mathbf{Z}, t; \psi, \nabla\psi, \lambda) \equiv \int_E \varrho_R \hat{\sigma}(\mathbf{X}, \Phi) dE + \lambda^{\nu} \gamma_{\nu}(\mathbf{Z}, t, \psi, \nabla\psi). \tag{7.35}$$

Eqs. (7.33) can be called the Lagrange's equations of the second kind with undertermined multipliers.

The solutions of special problems of analytical mechanics of elastic media can be found in [1], [2], [6], [7] and in the related paper.

REFERENCES

[1] C. Woźniak, Non-linear mechanics of constrained material continua, I, II, Arch. Mech., 26, 1 and 28, 2, Warszawa 1974.

[2] C. Woźniak, Constrained continuous media I-III, Bull. Acad. Polon. Sci., ser.sci. techn., 21, 3-4, 1973.

[3] C. Woźniak, On the non-standard continuum mechanics, I, II, Bull. Acad. Polon. Sci., ser. sci. techn., 24, 1, 1976.

[4] C. Woźniak, Global theorems and field equations of the non-standard continuum mechanics, Bull. Acad. Polon. Sci. ser. sci. techn., 24, 2, 1976.

[5] C. Woźniak, Non-standard approach to the theory of elasticity I, II. Bull. Acad. Polon. Sci. ser. sci. techn., 24, 5, 1976.

[6] C. Woźniak, On the analytical mechanics of slender bodies, Bull. Acad. Polon. Sci. ser. sci. techn., 23, 4, 1975.

[7] C. Woźniak, Bodies reinforced by thin discrete layers, Bull. Acad. Polon. Sci. ser. sci. techn., 23, 4, 1975.

WAVE PROPAGATION
IN FINITELY DEFORMED ELASTIC MATERIAL

Zbigniew Wesolowski
Institute of Fundamental Technological
Research, Warsaw, Swietokrzyska 21
Poland

1. BASIC EQUATIONS OF NON-LINEAR ELASTICITY

Introduce two fixed coordinate systems : the system $\{X^\alpha\}$ and the system $\{x^i\}$ The body in reference configuration B_R is referred to the system $\{X^\alpha\}$, and the body in actual configuration B is referred to the system $\{x^i\}$. The material point is identified by its position in B_R ; by three numbers X^α, $\alpha = 1, 2, 3$. Position of this point at time t is determined by three numbers x^i, $i = 1, 2, 3$, which are its coordinates in the system $\{x^i\}$. Motion of the body is described by the function

$$x^i = \xi^i(X^\alpha, t). \tag{1.1}$$

The deformation gradient x^i_α, velocity $\dot{x}^i$ and acceleration $\ddot{x}^i$ are defined by the relations

$$x^i_\alpha = \frac{\partial \xi^i}{\partial X^\alpha}, \tag{1.2}$$

$$\dot{x}^i = \frac{\partial \xi^i}{\partial t}, \tag{1.3}$$

(1.4) $$\ddot{x}^i = \frac{\partial^2 \xi^i}{\partial t^2} .$$

The state of stress at particle X^α is defined by the Piola-Kirchhoff stress tensor T_{Ri}^{α}. Consider the surface $\mathcal{S}_R$ dividing the neighbourhood of X^φ into two parts, and denote by N_α the unit normal to $\mathcal{S}_R$. The density t_i of force actin on $\mathcal{S}_R$ (per unit area) is given by the formula

(1.5) $$t_i = T_{Ri}^{\alpha} N_\alpha .$$

Denote by ϱ_R the mass density of B_R, and by b_i the body force density (per unit mass). The equation of motion is

(1.6) $$T_{Ri}^{\alpha}\Big\|_\alpha + \varrho_R b_i = \varrho_R \ddot{x}_i \quad ,$$

where

$$(\,.\,)\|_\alpha = (\,.\,)|_\alpha + (\,.\,)|_r \, x^r_\alpha .$$

The double vertical line denotes total covariant differentiation, and the single vertical line denotes partial covariant differentiation. If the systems x^i and X^α are Cartesian and the differentiated function depends on X^α only, then the total covariant differentiation reduces to partial differentiation, denoted by comma.

Denote by σ the internal energy per unit mass. For the elastic material σ is a function of the deformation gradient $x^i{}_\alpha$ and entropy η

(1.7) $$\sigma = \sigma\left(x^i_\alpha , \eta\right).$$

This relation is local both in time, and in space. If the material is inhomogeneous the function σ depends explicitly on X^α. For various material symmetries the function σ takes special forms. The Piola-Kirchhoff stress tensor T_{Ri}^{α} and the temperature T may be calculated from the formulae

(1.8) $$T_{Ri}^{\alpha} = \varrho_R \frac{\partial \sigma}{\partial x^i_\alpha} \quad ,$$

(1.9) $$T = \frac{\partial \sigma}{\partial \eta} .$$

For isotropic process η = const. and in (1.8) entropy is a parameter. If T = const. the process is isothermal. In this case by replacing the internal energy σ by free energy $\sigma - T\eta$ one obtains again eq. (1.8) with T as a parameter.

Substitute (1.8) into (1.6). The motion of the body is governed by the equation

$$A_{i\ k}^{\alpha\ \beta} x^k{}_\beta \|_\alpha + \varrho_R b_i = \varrho_R \ddot{x}_i \quad , \tag{1.10}$$

where

$$A_{i\ k}^{\alpha\ \beta} = \varrho_R \frac{\partial^2 \sigma}{\partial x^i{}_\alpha \partial x^k{}_\beta} \tag{1.11}$$

For reasonable material $A_{i\ k}^{\alpha\ \beta} A_\alpha A_\beta$ and $A_{i\ k}^{\alpha\ \beta} B^i B^k$ for each A_α and B^i are positive definite symmetric tensors.

In order to obtain the linearized equations of motion consider the motion

$$\overset{*}{x}{}^i = \xi^i(X^\alpha, t) + u^i(X^\alpha, t) \, , \tag{1.12}$$

that slightly differs from the motion (1.1). In accord with (1.2) there is

$$\overset{*}{x}{}^i{}_\alpha = x^i{}_\alpha + u^i{}_\alpha \, . \tag{1.13}$$

Expanding $\overset{*}{T}_{Ri}{}^\alpha$ into the Taylor series at $x^i{}_\alpha$ one obtains the following expression for the stress tensor

$$\overset{*}{T}{}_{Ri}^{\alpha} = T_{Ri}^{\alpha} + A_{i\ k}^{\alpha\ \beta} u^k \|_\beta \quad , \tag{1.14}$$

and the linearized equations of motion

$$\left(A_{i\ k}^{\alpha\ \beta} u^k \|_\beta\right)\Big\|_\alpha + \varrho_R b_i = \varrho_R \ddot{u}_i \tag{1.15}$$

The equations (1.10) and (1.15) are the basis for all the subsequent calculations concerning the waves.

2. DISCONTINUITY SURFACE

Consider the time-dependent surface $\mathcal{S}$ dividing B_R into two parts. The surface $\mathcal{S}$ is defined by either of the equations

$$X^\alpha = X^\alpha(M^1, M^2, t)\,, \tag{2.1}$$

$$t = \varphi(X^\alpha)\,, \tag{2.2}$$

where M^K, $K = 1, 2$ are surface coordinates. The vectors

$$X^\alpha_{,K} = \frac{\partial X^\alpha}{\partial M^K}$$

are tangent to $\mathcal{S}$, and the vector

$$N_\alpha = \frac{\varphi_{,\alpha}}{\sqrt{\varphi_{,\varrho}\varphi_{,}^{\varrho}}}\,, \tag{2.3}$$

is the unit normal to $\mathcal{S}$. Therefore

$$N_\alpha X^\alpha_{,K} = 0\,. \tag{2.4}$$

The vector

$$X^\alpha_{,t} = \frac{\partial X^\alpha}{\partial t}\,, \tag{2.5}$$

is the velocity of the point $M^K =$ const. situated on $\mathcal{S}$. This velocity depends of the coordinate system $\{M^K\}$. Substitute (2.1) into (2.2). Differentiating the resulting identity with respect to time we obtain the identity

$$\varphi_{,\alpha} X^\alpha_{,t} = 1 \tag{2.6}$$

From (2.3) and (2.6) it follows

$$U = X^\alpha_{,t} N_\alpha = \frac{1}{\sqrt{\varphi_{,\varrho}\varphi_{,}^{\varrho}}}\,. \tag{2.7}$$

The projection of $X^{\alpha}{}_{,t}$ on N_{α} is therefore independent of the surface coordinate system $\{M^K\}$; U is the propagation velocity of the surface $\mathcal{S}$. It is the velocity of $\mathcal{S}$ in the direction of its own normal N_{α} .

Consider arbitrary field $H(X^{\alpha},t)$; H stands for scalar, vector, tensor etc. By assumption at the points outside $\mathcal{S}$ the field H is continuous. On $\mathcal{S}$ there may be discontinuity of H . Denote the front and back sides of $\mathcal{S}$ by $\mathcal{S}^F$, $\mathcal{S}^B$ respectively. It follows

$$\begin{aligned} H &= H(X^{\alpha},t) && \text{in } B_R, \\ H &= H^F(M^K,t) && \text{on } \mathcal{S}^F, \\ H &= H^B(M^K,t) && \text{on } \mathcal{S}^B, \end{aligned} \tag{2.8}$$

Consider two points A and B on $\mathcal{S}$, and two points A^F and B^F in front of $\mathcal{S}$, close to the points A and B (Fig. 1). For fixed time t there hold the approximate relations

$$H(A)-H(B) = \frac{\partial H^F}{\partial M^K}\,\Delta M^K,$$

$$H(A^F)-H(B^F) = \left(\frac{\partial H}{\partial X^{\alpha}}\right)^F \Delta X^{\alpha}.$$

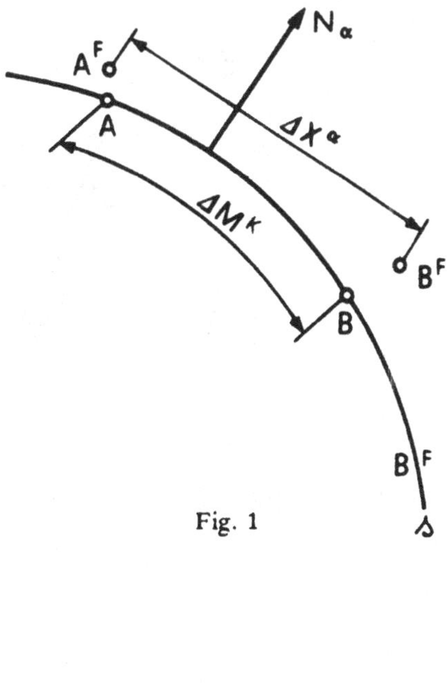

Fig. 1

For $A^F \rightarrow A$ and $B^F \rightarrow B$, by the continuity assumption there is $H(A^F) \rightarrow H(A)$ and $H(B^F) \rightarrow H(B)$. Therefore

$$\frac{\partial H^F}{\partial M^K} \Delta M^K = \left(\frac{\partial H}{\partial X^\alpha}\right)^F \Delta X^\alpha ,$$

and in the limit $\Delta X^\alpha \rightarrow 0$, $\Delta M^K \rightarrow 0$ we have

$$(2.9) \qquad \begin{aligned} H^F_{,K} &= (H_{,\alpha})^F X^\alpha_{,K} , \\ H^B_{,K} &= (H_{,\alpha})^B X^\alpha_{,K} . \end{aligned}$$

The equations $(2.9)_2$ follows from $(2.9)_1$ by changing the front side into the rear side. Equations (2.9) are the compatibility conditions for the discontinuity surface $\mathcal{S}$.

Consider in turn two positions $\mathcal{S}$ and $\mathcal{S}^*$ of the discontinuity surface $\mathcal{S}$, corresponding to time t and t^* (Fig. 2). Denote by A and A^* two points with the same surface coordinates M^K, situated on $\mathcal{S}$ and $\mathcal{S}^*$ respectively ; and by A^F and A^{*F} two

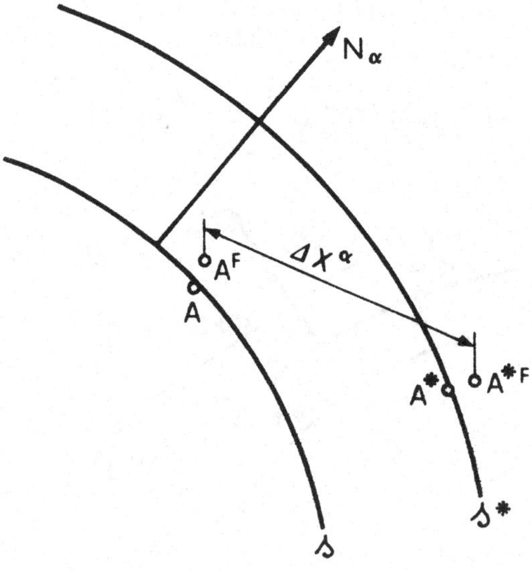

Fig. 2

points in front of $\mathcal{S}$ and $\mathcal{S}^*$, close to A and A^*. There hold the approximate relations

$$H^F(A^*) - H^F(A) = \frac{\partial H^F}{\partial t}(t^* - t),$$

$$H(A^{*F}) - H(A^F) = \left(\frac{\partial H}{\partial X^\alpha}\right)^F \Delta X^\alpha + \left(\frac{\partial H}{\partial t}\right)^F (t^* - t).$$

For $A^F \to A$ and $A^{*F} \to A$, by the continuity assumption there is $H(A^F) \to H(A)$ and $H(A^{*F}) \to H(A^*)$. Therefore

$$\frac{\partial H^F}{\partial t}(t^* - t) = \left(\frac{\partial H}{\partial X^\alpha}\right)^F \Delta X^\alpha + \left(\frac{\partial H}{\partial t}\right)^F (t^* - t),$$

and in the limit

$$\frac{\partial H^F}{\partial t} = \left(\frac{\partial H}{\partial t}\right)^F + \left(\frac{\partial H}{\partial X^\alpha}\right)^F X^\alpha_{,t},$$
$$\frac{\partial H^B}{\partial t} = \left(\frac{\partial H}{\partial t}\right)^B + \left(\frac{\partial H}{\partial X^\alpha}\right)^B X^\alpha_{,t}, \qquad (2.10)$$

Eq. $(2.10)_2$ follows from $(2.10)_1$ if we change the front side into the rear side. Equations (2.10) are the kinematical compatibility conditions.

Denote by the double bracket the jump of arbitrary field on $\mathcal{S}$

$$[\![\,.\,]\!] = (\,.\,)^B - (\,.\,)^F$$

Subtracting $(2.9)_1$ from $(2.9)_2$ and $(2.10)_1$ from $(2.10)_2$ we obtain

$$[\![H]\!]_{,K} = [\![H_{,\alpha}]\!] X^\alpha_{,K},$$
$$[\![H]\!]_{,t} = [\![H_{,t}]\!] + [\![H_{,\alpha}]\!] X^\alpha_{,t}. \qquad (2.11)$$

The jumps are functions of M^K and t, and are defined on the discontinuity surface $\mathcal{S}$. only. In general (2.1) may be solved for M^K, t; therefore the jumps may be expressed as the functions of X^α, $[\![\,.\,]\!] = f(X^\alpha)$. In the latter case the jump is given at the point X^α only for the instant t, when X^α is situated on $\mathcal{S}$.

Consider special case

$$[\![H]\!] = 0 \qquad (2.12)$$

The field $H(X^\alpha, t)$ (but not its derivatives) is continuous on $\mathcal{S}$. The left hand sides of (2.11) equal zero. In accord with (2.4) the equation $(2.11)_1$ leads to

(2.13) $$[\![H_{,\alpha}]\!] = AN_\alpha ,$$

where A is arbitrary parameter. From the equation $(2.11)_2$ we obtain

$$[\![H_{,t}]\!] = -AN_\alpha X^\alpha_{,t} ,$$

and using (2.7), finally

(2.14) $$[\![H_{,t}]\!] = -AU .$$

Because H may be taken as the physical quantity (i.e. displacement), or its integral, or its derivative, the compatibility conditions (2.13) and (2.14) are in fact as general, as the compatibility conditions (2.11). They are the basis for further considerations of weak and strong discontinuity waves.

3. ACCELERATION WAVE

Consider time-dependent discontinuity surface $\mathcal{S}$, on which the function $\xi^i(X^\alpha, t)$ (cf. (1.1)) and its first derivatives

$$x^i_\alpha , \dot{x}^i$$

are continuous. By assuption at least one of the second derivatives

$$x^i_{\alpha,\beta} , \dot{x}^i_{,\alpha} , \dot{x}^i_{,t} ,$$

is discontinuous.

Take in (2.13) and (2.14) first $H = x^i_\alpha$, and then $H = \dot{x}^i$. Both functions are continuous, therefore (2.13) and (2.14) may be used. We have

(3.1) $$[\![x^i_{\alpha,\beta}]\!] = A^i_\alpha N_\beta , \quad [\![x^i_{\alpha,t}]\!] = -A^i_\alpha U ,$$

(3.2) $$[\![\dot{x}^i_{,\alpha}]\!] = B^i N_\alpha , \quad [\![\dot{x}^i_{,t}]\!] = -B^i U ,$$

where A^i_α and B^i are parameters. Because $x^i_{\alpha,\beta}$ is symmetric in α , β there holds the relation

$$A^i_\alpha N_\beta = A^i_\beta N_\alpha$$

Multiplying both sides by N^β we get

$$A^i_\alpha = A^i_\beta N_\alpha N^\beta = A^i N_\alpha \ , \quad A^i = A^i_\beta N^\beta \ ,$$

where $A^i = A^i_\beta N^\beta$ denotes new set of parameters. Of course $\dot{x}^i_{,\alpha} = x^i_{\alpha,t}$. In accord with $(3.1)_2$ and $(3.2)_1$ we have

$$B^i N_\alpha = - A^i N_\alpha U \ ,$$

therefore

$$B^i = - A^i U \ .$$

Finally the jumps of second derivatives are

$$\begin{aligned} [\![x^i_{,\alpha\beta}]\!] &= A^i N_\alpha N_\beta \ , \\ [\![x^i_{,\alpha t}]\!] &= - A^i N_\alpha U \ , \\ [\![x^i_{,tt}]\!] &= A^i U^2 \ . \end{aligned} \tag{3.3}$$

By assumption the acceleration $\ddot{x}^i$ is discontinuous on $\mathcal{S}$. Such a discontinuity surface $\mathcal{S}$ is called the acceleration wave (acceleration $\ddot{x}^i$ is discontinuous on $\mathcal{S}$).

The covariant derivatives of a function differ from the partial derivatives only by products of Chistoffel symbols and the function itself. From this fact we infer

$$\begin{aligned} [\![x^i_\alpha \|_\beta]\!] &= A^i N_\alpha N_\beta \ , \\ [\![\dot{x}^i \|_\alpha]\!] &= - A^i N_\alpha U \ , \\ [\![\ddot{x}^i]\!] &= A^i U^2 \end{aligned} \tag{3.4}$$

because x^i_α and $\dot{x}^i$ are continuous. From the tensor character of the left-hand sides of (3.4) it follows that A^i is a vector. This vector determines the jumps of the second derivatives of x^i and is called the amplitude of the acceleration wave. The propagation speed of this wave is U .

In order to find the propagation condition note that (1.10) is satisfied on both sides of $\mathcal{S}$, therefore

$$\left[\!\left[A_{i\,k}^{\alpha\,\beta} x^k{}_\beta\|_\alpha \right]\!\right] + \left[\!\left[\varrho_R b_i \right]\!\right] = \left[\!\left[\varrho_R \ddot{x}_i \right]\!\right] . \tag{3.5}$$

The functions $A_{i\,k}^{\alpha\,\beta}$ depend on x^i_α and are continuous on $\mathcal{S}$. Assume that the density ϱ_R and body force b_i are continuous (the opposite case will be considered in chapter 5). From (3.5) it follows

$$A_{i\,k}^{\alpha\,\beta} \left[\!\left[x^k{}_\beta\|_\alpha \right]\!\right] = \varrho_R \left[\!\left[\ddot{x}_i \right]\!\right] . \tag{3.6}$$

Substituting the relations (3.4) into (3.6) we have finally

$$\left(Q_{ik} - \varrho_R U^2 g_{ik} \right) A^k = 0 \quad , \tag{3.7}$$

where

$$Q_{ik} = A_{i\,k}^{\alpha\,\beta} N_\alpha N_\beta \quad , \tag{3.8}$$

is the acoustic tensor for the propagation direction N_α . In accord with the propagation condition (3.7) the amplitude A^k is the proper vector, and the product $\varrho_R U^2$ is the proper value of the acoustic tensor Q_{ik} . From the symmetry $A_{i\,k}^{\alpha\,\beta} = A_{k\,i}^{\beta\,\alpha}$ it follows $Q_{ik} = Q_{ki}$. Because symmetric tensor possesses three mutually orthogonal proper vectors, therefore for given propagation direction there exist three mutually orthogonal possible amplitudes, Fig. 3.

The corresponding proper numbers $\varrho_R U^2$ are real. To the positive proper numbers there correspond positive speeds U and the wave may propagate. Both $+U$ und $-U$ are possible speeds, therefore acceleration wave propagating in the N_α direction may propagate (with the same amplitude) in the $-N_\alpha$ direction. In real materials all the proper numbers $\varrho_R U^2$ are positive, cf. remark after eq. (1.11).

The equation (3.8) determines the speed U and the direction of the amplitude

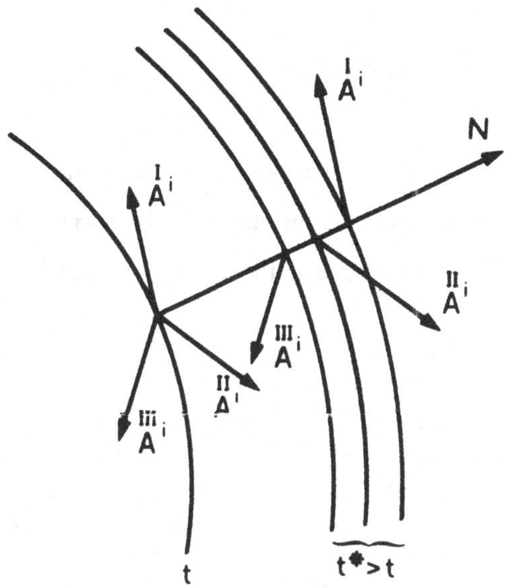

Fig. 3

A^k . The magnitude of A^k , measuring the intensity of the wave is governed by transport equation, which we shall derive in Chapter 6.

It must be stressed that in accord with the propagation condition (3.7) the propagation speed does not depend on the intensity of the wave. The function (1.1) maps the surface $\mathfrak{s}$ in B_R into the surface $\mathfrak{s}$ in B . Denote the unit normal to $\mathfrak{s}$ by n_i . If A^i is parallel to n_i the wave is longitudinal ; if A^i is orthogonal to n_i the wave is transverse. Typical wave is neither longitudinal, nor transverse.

Taking into account (2.3) and (2.7) we have $N_\alpha = U\varphi_{,\alpha}$. Therefore the equivalent form of propagation condition (3.7) is

$$\left(A_i{}^\alpha{}_k{}^\beta \varphi_{,\alpha} \varphi_{,\beta} - \varrho_R g_{ik}\right) A^k = 0 \ . \tag{3.9}$$

The propagation condition (3.9) (or (3.7)) is a system of homogeneous algebraic equations. The non-trivial solution exists iff

$$\det\left(Q_{ik} - \varrho_R U^2 g_{ik}\right) = 0 \ ,$$

or

(3.10) $$\det\left(A_{i\ k}^{\alpha\beta}\,\Psi_{,\alpha}\Psi_{,\beta}-\varrho_R\,g_{ik}\right)=0\ .$$

This non-linear first order differential equation together with initial conditions determines the wave front $t=\Psi(X^{\alpha})$. There are six solutions to (3.10), namely $\pm\overset{1}{\Psi},\pm\overset{2}{\Psi},\pm\overset{3}{\Psi}$. Denote $\xi_\alpha=\Psi_{,\alpha}$. The equation (3.10) determines in the space ξ_α a surface consisting of three sheets, Fig. 4.

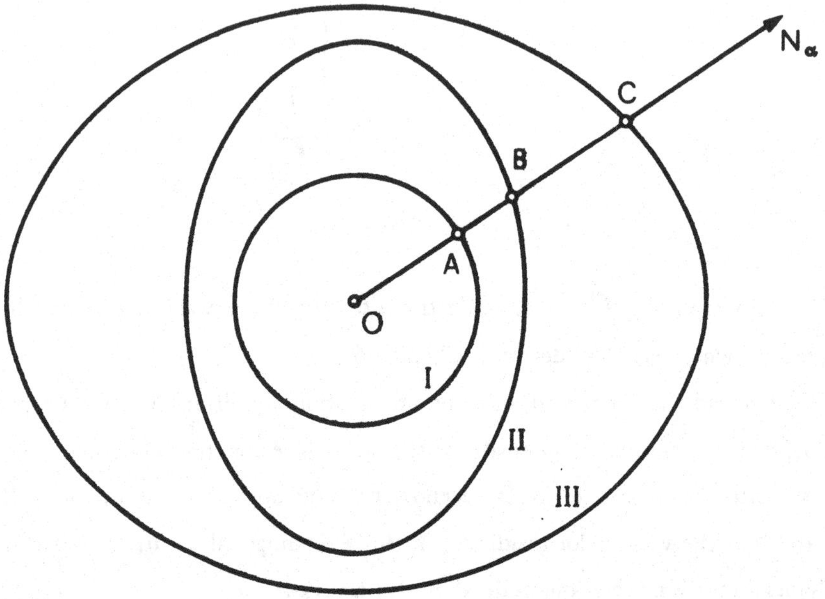

Fig. 4

Because of the relation $\xi_\alpha=N_\alpha/U$ the radius vector ξ_α has length equal to $1/U$. To each N_α there correspond in general three different speeds $\overset{1}{U}$, $\overset{2}{U}$, $\overset{3}{U}$. The surface (3.10), shown on Fig. 4 is the slowness surface.

4. ACOUSTIC WAVE

The surface $\mathcal{S}$ is the n-th order wave if the function (1.1) and its derivatives up to (n - 1) - th are continuous, and at least one of the n-th derivatives is discontinuous. Consider first the third order wave. Taking in (2.13) and (2.14) in turn $H = x^i_{\alpha,\beta}$, $H = \dot{x}^i_{\alpha}$ and $H = \dot{x}^i_{,t}$ we have

$$\left[\!\left[x^i_{\alpha,\beta\gamma} \right]\!\right] = A^i_{\alpha\beta} N_\gamma \ , \quad \left[\!\left[x^i_{\alpha,\beta t} \right]\!\right] = - A^i_{\alpha\beta} U \ ,$$
$$\left[\!\left[\dot{x}^i_{,\alpha\beta} \right]\!\right] = B^i_\alpha N_\beta \ , \quad \left[\!\left[\dot{x}^i_{,\alpha t} \right]\!\right] = - B^i_\alpha U \ , \tag{4.1}$$
$$\left[\!\left[\dot{x}^i_{,t\alpha} \right]\!\right] = C^i N_\alpha \ , \quad \left[\!\left[\dot{x}^i_{,tt} \right]\!\right] = - C^i U \ ,$$

where $A^i_{\alpha\beta}$, B^i_α , C^i , are parameters.

Calculations similar to that given in the previous chapter lead to the following expressions

$$\left[\!\left[x^i_\alpha \|_{\beta\gamma} \right]\!\right] = A^i N_\alpha N_\beta N_\gamma \ ,$$
$$\left[\!\left[\dot{x}^i \|_{\alpha\beta} \right]\!\right] = - A^i N_\alpha N_\beta U \ ,$$
$$\left[\!\left[\ddot{x}^i \|_{\alpha} \right]\!\right] = A^i N_\alpha U^2 \ , \tag{4.2}$$
$$\left[\!\left[\dddot{x}^i \right]\!\right] = - A^i U^3 \ ,$$

where A^i is a vector characterizing the amplitude of the third order wave (cf. (4.2)).

Differentiate (1.10) with respect to time. We get

$$A_{i\ k}^{\alpha\ \beta} \dot{x}^k \|_{\alpha\beta} + A_{i\ k\ m}^{\alpha\ \beta\ \gamma} x^k_\beta \|_\alpha \dot{x}^m \|_\gamma = \varrho_R \dddot{x}^i \ , \tag{4.3}$$

where

$$A_{\ k\ m}^{\alpha\ \beta\ \gamma} = \varrho_R \frac{\partial^3 \sigma}{\partial x^i_\alpha \, \partial x^k_\beta \, \partial x^m_\gamma} \ . \tag{4.4}$$

Equation (4.3) is satisfied on both sides of $\mathcal{S}$, therefore

$$\left[\!\left[A_{i\ k}^{\alpha\ \beta} \dot{x}^k \|_{\alpha\beta} \right]\!\right] + \left[\!\left[A_{i\ k\ m}^{\alpha\ \beta\ \gamma} x^k_\beta \|_\alpha \dot{x}^m \|_\gamma \right]\!\right] = \left[\!\left[\varrho_R \dddot{x}_i \right]\!\right] . \tag{4.5}$$

Because $\left[\!\left[x^i_\alpha \right]\!\right] = 0$ there is

$$\left[\!\left[A_{i\ k}^{\alpha\ \beta} \right]\!\right] = \left[\!\left[A_{i\ k\ m}^{\alpha\ \beta\ \gamma} \right]\!\right] = 0 \ ,$$

and (4.5) reduces to the equation

$$A_{i\,k}^{\alpha\beta}\left[\!\left[\dot{x}^{k}\|_{\alpha\beta}\right]\!\right] = \varrho_R\left[\!\left[\dddot{x}_i\right]\!\right] \tag{4.6}$$

Inserting the expressions (4.2) into (4.6) we obtain the propagation condition

$$\left(Q_{ik} - \varrho_R\, g_{ik}\right) A^k N_\gamma = 0 \;, \tag{4.7}$$

where

$$Q_{ik} = A_{i\,k}^{\alpha\beta} N_\alpha N_\beta \,.$$

Because N_γ is the unit vector the coefficient of N_γ in (4.7) must vanish. Finally the propagation condition for the third order wave is

$$\left(Q_{ik} - \varrho_R\, g_{ik}\right) A^k = 0 \,. \tag{4.8}$$

Comparing this equation with (3.7) we infer that the third order wave propagates as the acceleration wave.

Consider in turn n-th wave. Differentiating (1.10) (n - 2) — times with respect to time we obtain

$$A_{i\,k}^{\alpha\beta}\, x^{k}_{\beta}\|_{\alpha,\underbrace{t\ldots t}_{n-2}} + \ldots = \varrho_R\, \ddot{x}_{i,\underbrace{t\ldots t}_{n-2}} \tag{4.9}$$

Taking in (2.13) and (2.14)

$$H = x^{i}_{\alpha,\underbrace{\beta\ldots\delta}_{n-1}} \;, \qquad H = \dot{x}^{i}_{,\underbrace{t\ldots t}_{n-1}}$$

We arrive to the relations similar to (3.1), and finally to the expression

$$\begin{aligned} \left[\!\left[x^{i}_{\alpha,\beta\underbrace{t\ldots t}_{n-2}}\right]\!\right] &= A^i U^{n-2} N_\alpha N_\beta \;, \\ \left[\!\left[x^{i}_{,\underbrace{tt\ldots t}_{n}}\right]\!\right] &= A^i U^n \,. \end{aligned} \tag{4.10}$$

After calculating the jump of the equation (4.9) and substituting (4.10) we arrive again to the propagation condition (3.7) (or (4.8)). It follows that all the waves of order $n \geqslant 2$ propagate with the same velocity U. Superposition of all waves of order $n \geqslant 2$ is the acoustic wave (or sound wave). The propagation speed of acoustic wave equals U Because of this fact U is called the sound speed.

5. REFLECTION AND REFRACTION OF ACCELERATION WAVE

Introduce Catesian coordinate system and denote $\xi_\alpha = \varphi_{,\alpha}$. In accord with (3.10) there is

$$N = \det\left(A_{i\,k}^{\alpha\beta}\,\xi_\alpha \xi_\beta - \varrho_R g_{ik}\right) = 0 \,. \tag{5.1}$$

The equation (5.1) defines the slowness surface, Fig. 4 (shown the intersection with plane $\xi^3 = 0$). The distances OA, OB, OC are equal to $1/\overset{I}{U}, 1/\overset{II}{U}, 1/\overset{III}{U}$, $\overset{I}{U} > \overset{II}{U} > \overset{III}{U}$. gation direction there correspond three vectors

$$\overset{I}{\xi}_\alpha = N_\alpha / \overset{I}{U}\,,\quad \overset{II}{\xi}_\alpha = N_\alpha / \overset{II}{U}\,,\quad \overset{III}{\xi}_\alpha = N_\alpha / \overset{III}{U}\,. \tag{5.2}$$

The wave surface is the wave front at time $t = 1$ produced by disturbance situated at $\xi^\alpha = 0$, acting at the instant $t = 0$. From this definition it follows that wave surface is the envelope of plane acceleration waves $N_\alpha \xi^\alpha = Ut$ at time $t = 1$, that at time $t = 0$ were passing the point $\xi^\alpha = 0$, Fig. 5.

Denote the coordinates of the point C (Fig. 5) by χ^α. It follows

$$\chi^\alpha \frac{\xi^\alpha}{\sqrt{\xi_\varrho \xi^\varrho}} = U \,,$$

$$\chi^\alpha \xi_\alpha = U \sqrt{\xi_\varrho \xi^\varrho} = 1 \,.$$

The plane wave at $t = 1$, shown at Fig. 5 is therefore given by the equation

$$\xi_\alpha X^\alpha = 1 \tag{5.3}$$

The vector ξ_α is orthogonal to this plane. Write the equation of the wave surface in the form

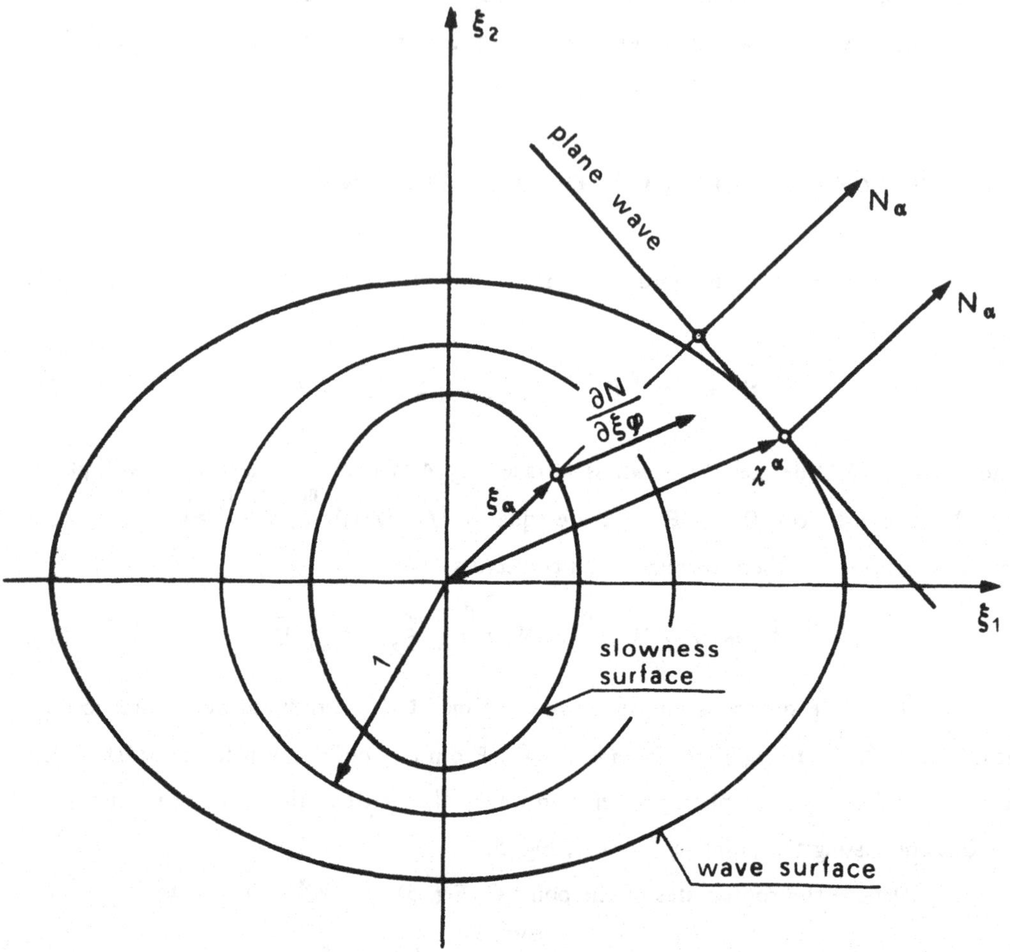

$$|\xi_\alpha| = \frac{1}{U(N_\alpha)}$$

$$|\chi^\alpha| = \frac{1}{|\xi_\alpha|} = U(N_\alpha)$$

Fig. 5

$$\chi^{\alpha} = \chi^{\alpha}(\xi_{\beta}) .$$

It follows that the vector

$$\frac{\partial \chi^{\alpha}(\xi_{\beta})}{\partial \xi_{\varphi}} \, d\xi_{\varphi} ,$$

for arbitrary $d\xi_{\varphi}$ is tangent to the wave surface. Because the plane $\xi_{\alpha} X^{\alpha} = 1$ with normal ξ_{α} is tangent to this surface there holds the orthogonality condition

$$\left(\frac{\partial \chi^{\alpha}(\xi_{\beta})}{\partial \xi_{\varphi}} \, d\xi_{\varphi}\right) \xi_{\alpha} = 0 .$$

On the other hand from (5.1) it follows that for arbitrary $d\xi_{\varphi}$ there is

$$\frac{\partial N(\xi_{\beta})}{\partial \xi_{\varphi}} \, d\xi_{\varphi} = 0 .$$

Comparing the last two relations we obtain

$$\frac{\partial \chi^{\alpha}(\xi_{\beta})}{\partial \xi_{\varphi}} \, \xi_{\alpha} = \nu \, \frac{\partial N(\xi_{\beta})}{\partial \xi_{\varphi}} . \tag{5.4}$$

Take now into account the relation (5.3), namely the relation

$$\chi^{\alpha} \, \xi_{\alpha} = 1 .$$

Differentiating it with respect to ξ_{β} we get

$$\frac{\partial \chi^{\alpha}(\xi_{\beta})}{\partial \xi_{\beta}} \, \xi_{\alpha} + \chi^{\alpha}(\xi_{\beta}) \, \delta_{\alpha}^{\beta} = 0 .$$

In accord with (5.4) we get finally

$$\chi^{\beta}(\xi_{\alpha}) = - \nu \, \frac{\partial N(\xi_{\alpha})}{\partial \xi_{\beta}} . \tag{5.5}$$

The vector $\chi^{\beta}(\xi_\alpha)$ is therefore parallel to the vector orthogonal to the slowness surface at ξ_α. Note that the vector ξ_α is parallel to the vector N_α orthogonal to the wave surface at χ^{β} (cf. Fig. 5). From this property and the equation (5.3) follows easy construction of the wave surface, if the slowness surface is known, and the construction of the slowness surface, if the wave surface is known. Two surfaces adjoined to each other in the manner described above are the polar reciprocal surfaces.

Consider now the reflected and refracted waves. Assume that the fixed surface Γ dividing two media I and II is given by either of the equations

$$X^\alpha = H^\alpha(M^1, M^2), \quad \Gamma(X^\alpha) = 0. \tag{5.6}$$

The surface parameters are denoted by M^1 and M^2. The incident we front is given by the equation

$$t = \overset{i}{\Phi}(X^\alpha) \tag{5.7}$$

At time t this front intersects Γ producing the curve l defined by the equation

$$t = \overset{i}{\Phi}\left(H^\alpha(M^1, M^2)\right) = F(M^1, M^2). \tag{5.8}$$

The incident wave produces the reflected wave

$$t = \overset{r}{\Phi}(X^\alpha), \tag{5.9}$$

and the refracted wave (t stands for "transmitted")

$$t = \overset{t}{\Phi}(X^\alpha). \tag{5.10}$$

Both waves intersect the surface Γ at the curve l, therefore

$$\overset{r}{\Phi}\left(H^\alpha(M^1, M^2)\right) = \overset{t}{\Phi}\left(H^\alpha(M^1, M^2)\right) = F(M^1, M^2). \tag{5.11}$$

Differentiate $(5.11)_1$ with respect to the surface parameters M^1 and M^2. Denoting $\overset{i}{\xi}_\alpha = \overset{i}{\Psi}_{,\alpha}$, $\overset{r}{\xi}_\alpha = \overset{r}{\Psi}_{,\alpha}$ we have

$$\left(\overset{i}{\xi}_1 - \overset{r}{\xi}_1\right)H^1_{,K} + \left(\overset{i}{\xi}_2 - \overset{r}{\xi}_2\right)H^2_{,K} + \left(\overset{i}{\xi}_3 - \overset{r}{\xi}_3\right)H^3_{,K} = 0 \tag{5.12}$$

Both waves propagate in the medium I, therefore denoting by $\overset{I}{A}{}_i{}^\alpha{}_k{}^\beta$ the value of $A_i{}^\alpha{}_k{}^\beta$ for the medium I there hold the equations

$$\det\left(\overset{I}{A}{}_i{}^\alpha{}_k{}^\beta \overset{i}{\xi}_\alpha \overset{i}{\xi}_\beta - \varrho_R g_{ik}\right) = 0,$$
$$\det\left(\overset{I}{A}{}_i{}^\alpha{}_k{}^\beta \overset{r}{\xi}_\alpha \overset{r}{\xi}_\beta - \varrho_R g_{ik}\right) = 0. \tag{5.13}$$

In accord with (5.12) the vector $\overset{i}{\xi}_\alpha - \overset{r}{\xi}_\alpha$ is orthogonal to the vectors $H^\alpha_{,K}$ which are tangent to the surface Γ dividing the media I and II. Therefore $\overset{i}{\xi}_\alpha - \overset{r}{\xi}_\alpha$ has the direction of the vector $K_\alpha = \Gamma_{,\alpha}$ orthogonal to Γ. Equations (5.12) define a prime p parallel to K_α. Because the point $\overset{i}{\xi}_\alpha$ is situated on p the parametric equation of p is

$$\xi_\alpha = \overset{i}{\xi}_\alpha + p K_\alpha, \tag{5.14}$$

where p is the parameter.
Substituting (5.14) into $(5.13)_2$ we obtain algebraic equation of sixth degree. There are at most six intersection points of this prime with the slowness surface, Fig. 6.

Heavy lines are the intersection lines of the slowness surface (three sheets) with the plane spanned by vectors $\overset{i}{\xi}_\alpha$ and K_α. Three of the intersection points of p with the slowness surface do not correspond to the reflected waves, because reflected waves must run behind the incident wave. The waves corresponding to the points A_1 and A_2 run in front of the incident wave, and therefore must be excluded. This is evident from the Fig. 7 (corresponding to the Fig. 6).

The dashed lines are the wave fronts corresponding to $A_1, A_2, \ldots, A_6$. If π is a plane orthogonal to K_α passing by $\xi_\alpha = 0$, then $\overset{i}{\xi}_\alpha$ is situated on one side of π and $\overset{r}{\xi}{}_\alpha^{(1)}$, $\overset{r}{\xi}{}_\alpha^{(2)}$, $\overset{r}{\xi}{}_\alpha^{(3)}$ are situated on the opposite side of π. In special cases there may exist two, or one reflected wave only, Fig. 8a, b, or no reflected wave at all, Fig. 8c.

If there may exist three reflected waves (as on Fig. 6) the reflection is regular. In order to explain the situation non-regular reflection consider slowness surface, that consists of three spheres (isotropic case). Assume that at $t = 0, 1, 2, 3, \ldots$ the incident wave reaches the points $A_0, A_1, A_2, \ldots$, Fig. 9b. The incident wave produces at time t a

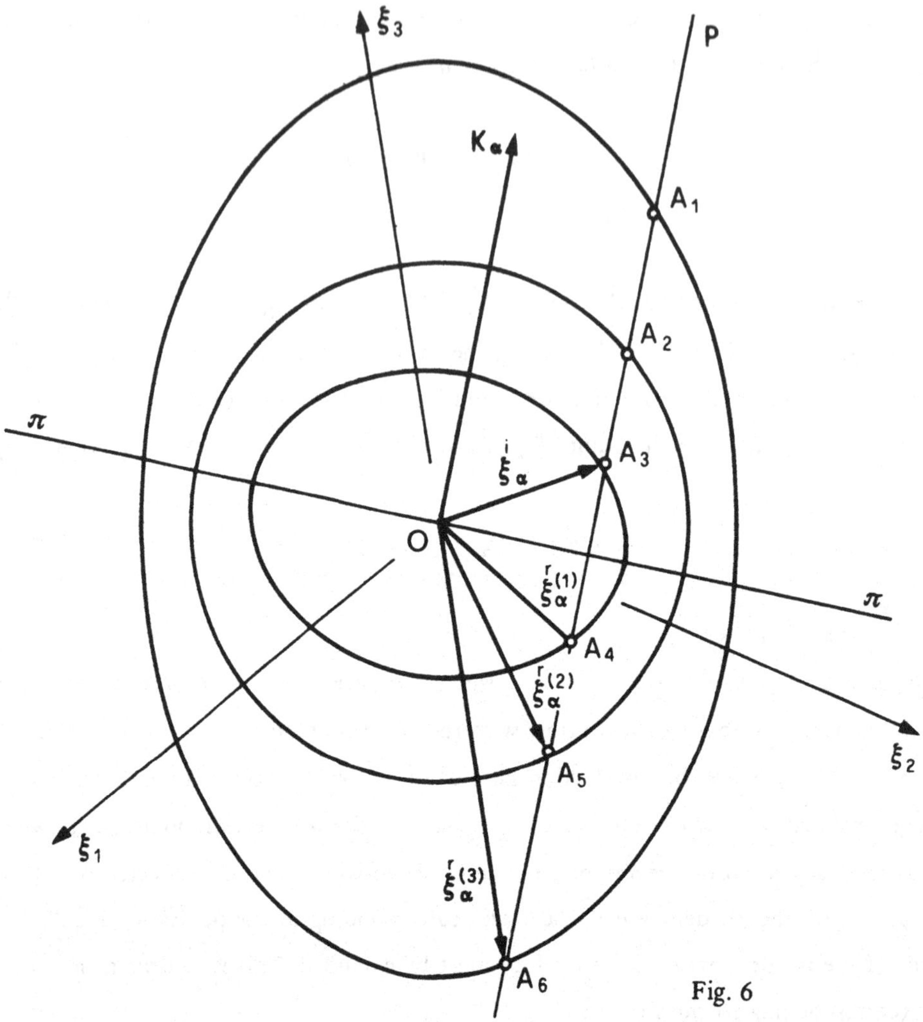

Fig. 6

disturbance at the point A_t. This disturbance propagates in the medium I. The front of the reflected wave is the envelope of the fronts of this waves. On the Fig. 9b heavy lines show the fronts of the incident wave at $t = 0, 1, 2, \ldots,$ and light lines the fronts of the disturbances at $t = 4$. The dashed lines are the fronts of the reflected waves. If there exist less than three reflected waves the situation is quite different. On the Fig. 10 is shown the case, when there exist two reflected waves only. Both are constructed in the same way, as on Fig. 9. The third wave front travels faster along Γ, than the incident wave front (at $t = 4$ reaches the point B lying in front of the incident wave). It is seen from Fig. 8

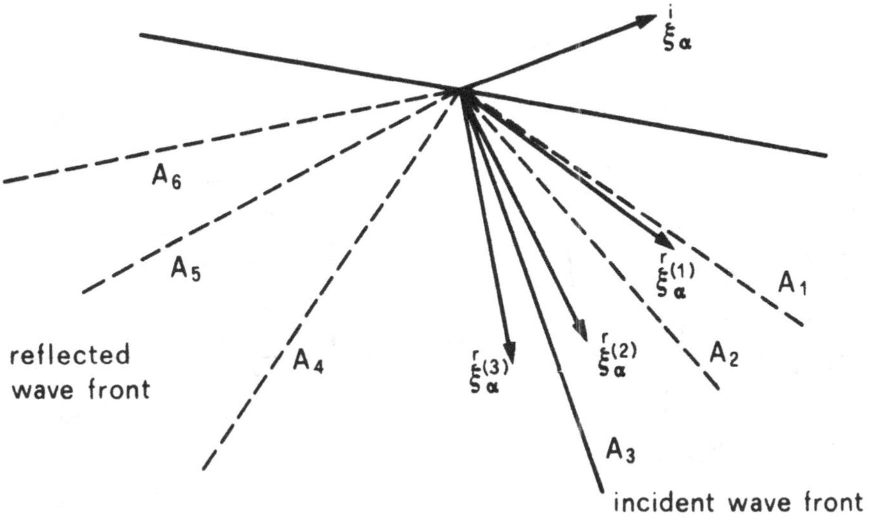

Fig. 7

that in this case the vector $\overset{r}{\xi}_\alpha$ is complex (not real). The wave with complex ξ_α is the surface wave.

Pass to the refracted waves. Differentiating $(5.11)_2$ with respect to M^1 and M^2 we get

$$\left(\overset{i}{\xi}_\alpha - \overset{t}{\xi}_\alpha\right) H^\alpha_{,1} = 0 ,$$
$$\left(\overset{i}{\xi}_\alpha - \overset{t}{\xi}_\alpha\right) H^\alpha_{,2} = 0 . \tag{5.16}$$

It follows that $\overset{i}{\xi}_\alpha - \overset{t}{\xi}_\alpha$ has the direction of the vector K_α orthogonal to Γ . Equations (5.15) define the prime line passing by $\overset{i}{\xi}_\alpha$ and parallel to K_α , the same as for reflected waves. All the vectors $\overset{i}{\xi}_\alpha$, $\overset{r}{\xi}_\alpha$ and $\overset{t}{\xi}_\alpha$ are therefore coplanar.

The incident wave propagates in the medium I, and the refracted waves in the medium II. In accord with (5.1) we have

$$\det\left(\overset{I}{A}{}^{\alpha\beta}_{i\,k}\, \overset{i}{\xi}_\alpha \overset{i}{\xi}_\beta - \varrho_R\, g_{ik}\right) = 0 ,$$
$$\det\left(\overset{II}{A}{}^{\alpha\beta}_{i\,k}\, \overset{t}{\xi}_\alpha \overset{t}{\xi}_\beta - \varrho_R\, g_{ik}\right) = 0 . \tag{5.15}$$

There are three sheets of the slowness surface for the medium I, and there sheets

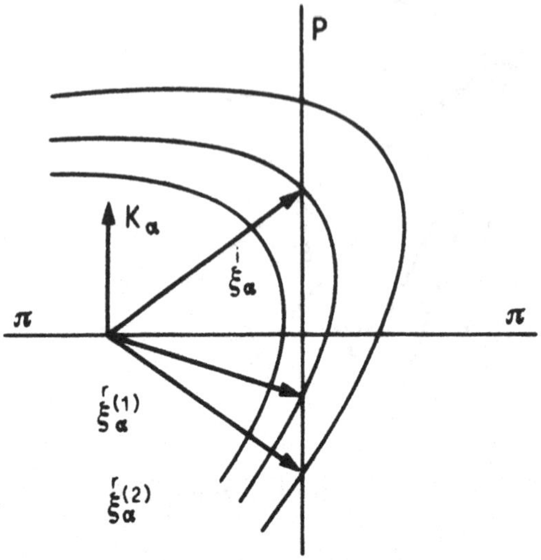

Fig. 8a

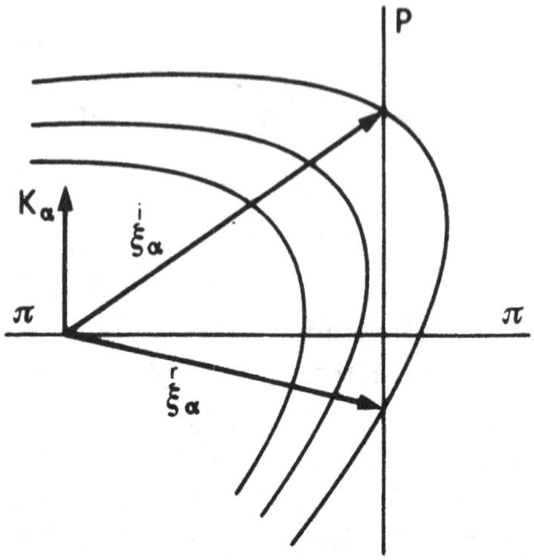

Fig. 8b

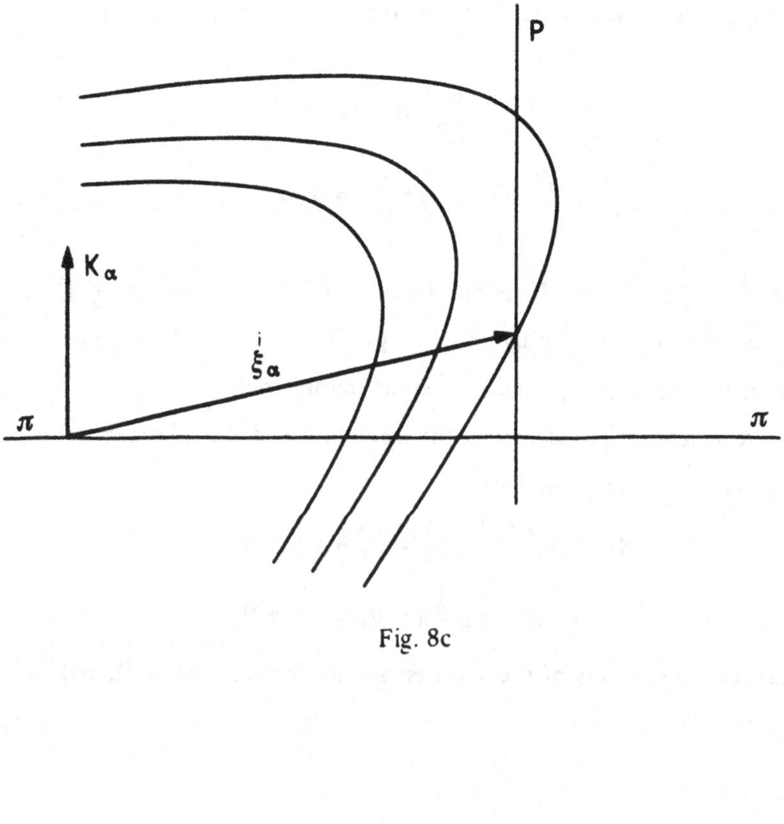

Fig. 8c

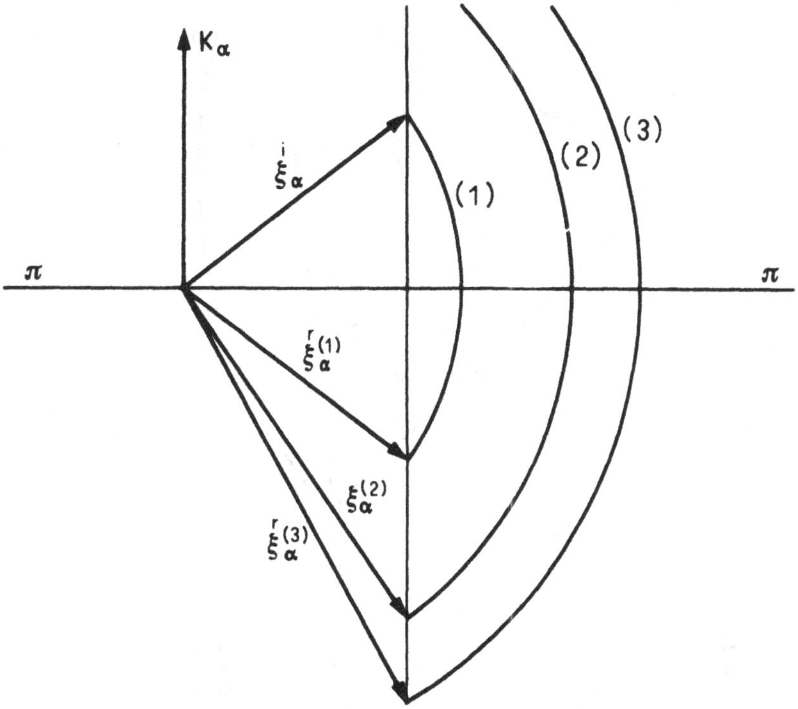

K_α
$\overset{i}{\xi}_\alpha$
(1)
(2)
(3)
π
π
$\overset{r}{\xi}{}^{(1)}_\alpha$
$\xi^{(2)}_\alpha$
$\overset{r}{\xi}{}^{(3)}_\alpha$

Fig. 9a

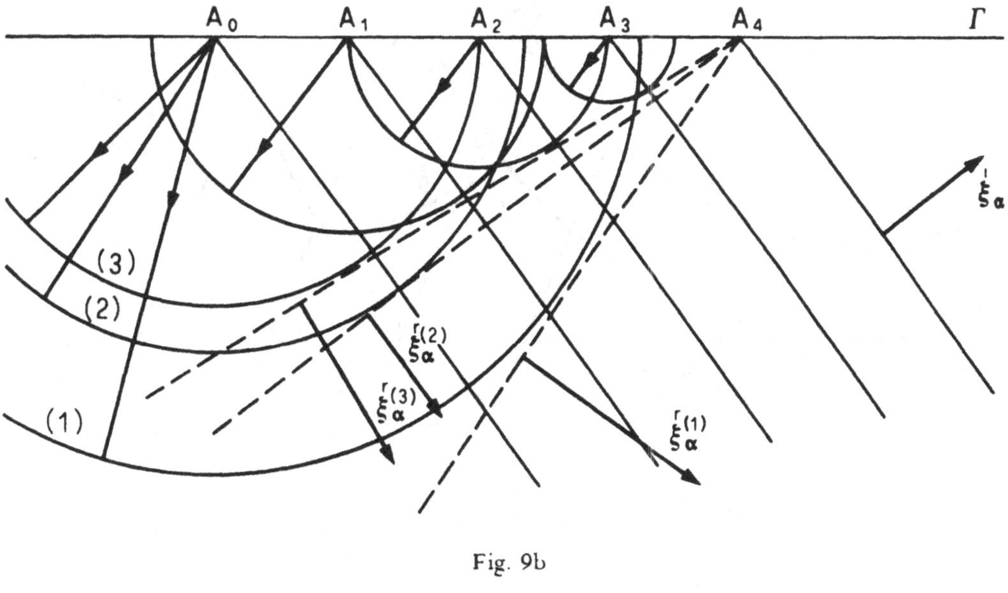

A_0
A_1
A_2
A_3
A_4
Γ
$\overset{i}{\xi}_\alpha$
(3)
(2)
(1)
$\overset{r}{\xi}{}^{(2)}_\alpha$
$\overset{r}{\xi}{}^{(3)}_\alpha$
$\overset{r}{\xi}{}^{(1)}_\alpha$

Fig. 9b

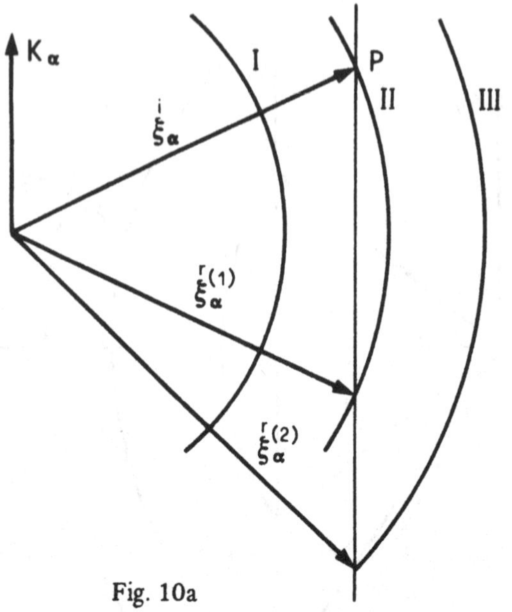

Fig. 10a

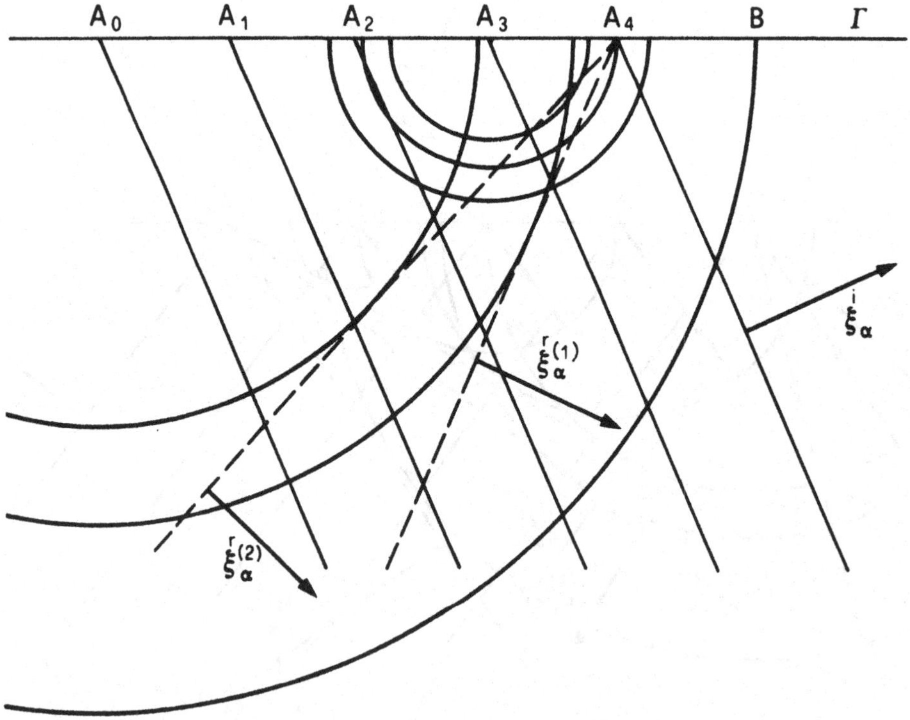

Fig. 10b

for the slowness surface for the medium II, Fig. 11. Heavy lines correspond to the medium I, and dashed lines to the medium II. The plane π is orthogonal to K_α . The refracted waves are produced on Γ , therefore the wave vectors $\overset{t}{\xi}_\alpha$ of the refracted waves and the wave vector $\overset{i}{\xi}_\alpha$ of the incident wave are situated at the same side of π . It is seen that in general there exist three reflected, and three refracted waves.

6. TRANSPORT EQUATION

Confine now to the Cartesian coordinates and consider the linearized equations of motion (1.15). Assuming that u^i , $u^i_{,\alpha}$, $\dot{u}^i$ are continuous on $\mathcal{S}$ we have (similary as for $x^i(X^\alpha, t)$, cf. (2.13), (2.14), (3.4))

$$\begin{aligned} \left[\!\left[u^i\|_{\alpha\beta} \right]\!\right] &= A^i N_\alpha N_\beta \, , \\ \left[\!\left[\dot{u}^i\|_\alpha \right]\!\right] &= -A^i N_\alpha U \, , \\ \left[\!\left[\ddot{u}^i \right]\!\right] &= A^i U^2 . \end{aligned} \tag{6.1}$$

Substituting this relations into (1.15) and taking into account that $u^i\|_\alpha$ and $\dot{u}^i$ are continuous we arrive again to the propagation condition (3.9), namely to the condition

$$\left(A_i{}^\alpha{}_k{}^\beta \varphi_\alpha \varphi_\beta - \varrho_R \delta_{ik} \right) A^k = 0 \, , \quad \varphi_\alpha = \varphi_{,\alpha} \, . \tag{6.2}$$

Assume that this equation has been solved and $\varphi(X^\alpha)$, $A^k(X^\alpha)$ are known (by assumption A^k is unit vector ; the actual amplitude is proportional to A^k). Eq. (6.2) concerns the discontinuity surface only. In order to determine the displacements behing the surface $\mathcal{S}$ represent $u^i(X^\alpha, t)$ in form of the series

$$u^i = \sum_{\nu=0}^{\infty} S_{\nu+2}(\Phi) g^i_\nu (X^\alpha, t) \, , \tag{6.3}$$

where

$$\begin{aligned} S_\nu &= \frac{1}{\nu!} \left[\frac{1}{2} (\Phi + |\Phi|) \right]^\nu , \\ \Phi(X^\alpha, t) &= \varphi(X^\alpha) - t \, . \end{aligned} \tag{6.4}$$

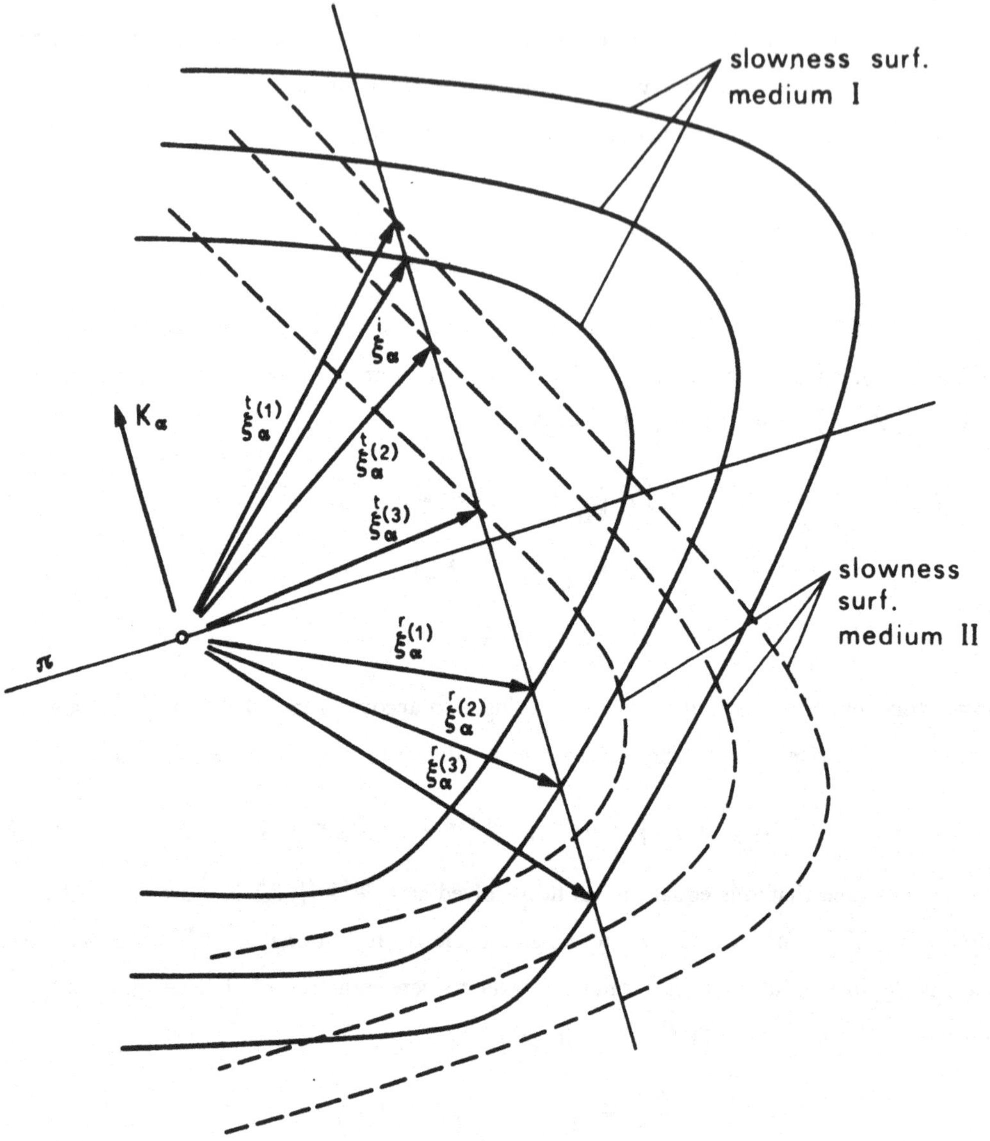
slowness surf.
medium I
slowness
surf.
medium II
K_α
$\overset{i}{\xi}_\alpha$
$\overset{t}{\xi}{}^{(1)}_\alpha$
$\overset{t}{\xi}{}^{(2)}_\alpha$
$\overset{t}{\xi}{}^{(3)}_\alpha$
$\overset{r}{\xi}{}^{(1)}_\alpha$
$\overset{r}{\xi}{}^{(2)}_\alpha$
$\overset{r}{\xi}{}^{(3)}_\alpha$
π

Fig. 11

It follows from $(6.4)_1$ that there holds the reccursive formulae

$$\frac{dS_\nu}{d\Phi} = S_{\nu-1} \;, \tag{6.5}$$

$$S_0 = \frac{dS_1}{d\Phi} = \eta(\Phi) \;, \tag{6.6}$$

where $\eta(\Phi)$ is the Heaviside function. On the discontinuity surface $\mathcal{S}$ there is $\Phi = 0$. It is seen from (6.3) that u^i , $u^i_{,\alpha}$, $\dot{u}^i$ are continuous on $\mathcal{S}$. The discontinuous derivatives of u^i are $u^i_{,\alpha\beta}$, $\ddot{u}^i$ (the second and higher order derivatives). In particular

$$\left[\!\left[u^i_{,\alpha\beta} \right]\!\right] = -S_0\,\Phi_\alpha\,\Phi_\beta\,g^i_0 \;, \qquad \left[\!\left[u^i_{,tt} \right]\!\right] = -S_0\,g^i_0 \;.$$

Note that

$$\Phi_\alpha = \Phi_{,\alpha} = \varphi_{,\alpha} = N_\alpha / U \;.$$

In accord with (6.1) the actual amplitude is therefore

$$-g^i_0 / U^2$$

Substituting (6.3) into (1.15) we arrive at the equation of the form

$$\sum_{\nu=0}^{\infty} S_\nu\,B^i_\nu = 0 \;,$$

where coefficients B^i_ν are differential expressions of g^i_μ . This equation is satisfied if all the coefficients B^i_ν are equal to zero (this is sufficient, but not necessary condition). The subsequent calculations will show, that taking $B^i_\nu = 0$ we can in fact find the solution. After calculations we write instead of $B^i_\nu = 0$ the system of equations

$$\left(A_i{}^\alpha{}_k{}^\beta\,\Phi_\alpha\,\Phi_\beta - \varrho_R\,\delta_{ik}\right) g^k_0 = 0 \;, \tag{6.7}$$

$$\left(A_i{}^\alpha{}_k{}^\beta\,\Phi_\alpha\,\Phi_\beta - \varrho_R\,\delta_{ik}\right) g^k_1 + \left(A_i{}^\alpha{}_k{}^\beta\,\Phi_\alpha\,g^k_{0,\beta} + A_i{}^\alpha{}_k{}^\beta\,\Phi_\beta\,g^k_{0,\alpha} + \right.$$

$$\left. + 2\varrho_R\,\delta_{ik}\,g^k_{0,t}\right) + \left(A_i{}^\alpha{}_k{}^\beta\,\Phi_{\alpha,\beta} + A_i{}^\alpha{}_{k}{}^\beta{}_{,\alpha}\,\Phi_\beta\right) g^k_0 = 0 \;, \tag{6.8}$$

$$\left(A_{i\,k}^{\alpha\beta}\Phi_\alpha\Phi_\beta-\varrho_R\delta_{ik}\right)g_{\nu+2}^{k}+\left(A_{i\,k}^{\alpha\beta}\Phi_\beta g_{\nu+1,\alpha}^{k}+\right.$$

$$\left.+A_{i\,k}^{\alpha\beta}\Phi_\alpha g_{\nu+1,\beta}^{k}+2\varrho_R\delta_{ik}g_{\nu+1,t}^{k}\right)+\left(A_{i\,k}^{\alpha\beta}\Phi_{\alpha,\beta}+\right.$$

(6.9) $$\left.+A_{i\,k}^{\alpha\beta}\Phi_\beta\right)g_{\nu+1}^{k}+\mathscr{L}_{i\gamma}g_\nu^\gamma=0\ ,\qquad \nu=0,1,2,\ldots$$

The only unknowns are g_0^k, g_1^k, ... , . Assume that all the proper numbers of the acoustic tensor Q_{ik} (and the tensor $A_i{}^\alpha{}_k{}^\beta\Phi_\alpha\Phi_\beta$) are different. If two, or three proper numbers coincide the analysis is similar, but more laborious. By comparing Eq. (6.7) with (3.9) we infer that

(6.10) $$g_0^k=\varkappa_0 A^k,$$

where $\varkappa_0$ is scalar coefficient. In order to find $\varkappa_0$ multiply (6.8) by A^i . The first term equals zero because of (6.7) and the symmetry $A_{i\,k}^{\alpha\beta}=A_{k\,i}^{\beta\alpha}$. Substitute into the remaining terms the expression (6.10). This leads to the partial differential equation

$$\left(A_{i\,k}^{\alpha\beta}\Phi_\alpha\varkappa_{0,\beta}+A_{i\,k}^{\alpha\beta}\Phi_\beta\varkappa_{0,\alpha}+2\varrho_R\delta_{ik}\varkappa_{0,t}\right)A^iA^k+$$

$$+\varkappa_0\left[A_{i\,k}^{\alpha\beta}\Phi_\alpha A^k_{,\beta}+A_{i\,k}^{\alpha\beta}\Phi_\beta A^k_{,\alpha}+2\varrho_R\delta_{ik}A^k_{,t}+\right.$$

(6.11) $$\left.+\left(A_{i\,k}^{\alpha\beta}\Phi_{\alpha,\beta}+A_{i\,k,\alpha}^{\alpha\beta}\Phi_\beta\right)A^k\right]A^i=0\ .$$

Introduce in the four-dimensional space $X^\alpha\times t$ the curve $\{b\}$ defined by the differential relations

(6.12)
$$\frac{dX^\alpha}{d\nu}=\left(A_i{}^\alpha{}_k{}^\beta\Phi_\beta+A_i{}^\beta{}_k{}^\alpha\Phi_\beta\right)A^iA^k,$$

$$\frac{dt}{d\nu}=2\varrho_R\delta_{ik}A^iA^k,$$

and initial conditions

$$t(\nu_0)=\varphi\left(X^\alpha(\nu_0)\right),$$

where ν is a parameter along the curve $\{b\}$. Note that in accord to $(6.4)_2$ the vector orthogonal to the surface $\Phi = 0$ (in the four-dimensional space $X^\alpha \times t$) has the coordinates

$$\left(\Phi_{,1}, \Phi_{,2}, \Phi_{,3}, \Phi_{,t}\right) = \left(\varphi_1, \varphi_2, \varphi_3, -1\right).$$

The scalar product of this vector and the vector tangent to $\{b\}$, defined by (6.12) is

$$2\left(A_{i\,k}^{\alpha\beta}\varphi_\alpha\varphi_\beta - \varrho_R\delta_{ik}\right)A^iA^k.$$

In accord with (3.9) (or (6.7)) this expression equals zero. By assumption one point of $\{b\}$ is situated on $\Phi = 0$. The above result shows that the whole curve $\{b\}$ is situated on $\Phi = 0$.

Calculate $dx_0/d\nu$. In accord with (6.12) we have

$$\frac{dx_0}{d\nu} = \frac{\partial x_0}{\partial X^\alpha}\frac{dX^\alpha}{d\nu} + \frac{\partial x_0}{\partial t}\frac{dt}{d\nu} =$$

$$= \left(A_{i\,k}^{\alpha\beta}\Phi_\alpha x_{0,\beta} + A_{i\,k}^{\alpha\beta}\Phi_\beta x_{0,\alpha} + 2\varrho_R\delta_{ik}x_{0,t}\right)A^iA^k.$$

It follows that the first term of (6.11) equals $dx_0/d\nu$. On the curve $\{b\}$ the coefficient of x_0 in (6.11) is a function of ν, because on this curve $X^\alpha = X^\alpha(\nu)$, $t = t(\nu)$. Denoting

$$P(\nu) = \Big[A_{i\,k}^{\alpha\beta}\Phi_\alpha A^k_{,\beta} + A_{i\,k}^{\beta\alpha}\Phi_\beta A^k_{,\alpha} +$$
$$+ 2\varrho_R\delta_{ik}A^k_{,t} + \left(A_{i\,k}^{\alpha\beta}\Phi_{\alpha,\beta} + A_{i\,k,\alpha}^{\alpha\beta}\Phi_\beta\right)A^k\Big]A^i \tag{6.13}$$

we can therefore represent Eq. (6.11) in form of ordinary differential equation along the curve $\{b\}$, namely

$$\frac{dx_0}{d\nu} + x_0 P(\nu) = 0.$$

Solution of this equation is the function

$$x_0 = C_0 \exp\left(-\int_0^\nu P(\nu)\,d\nu\right), \tag{6.14}$$

where C_0 is a constant.

Denote by $\{r\}$ the projection of the curve $\{b\}$ on the space X curve

$\{r\}$ is determined by $(6.12)_1$, Fig. 12. From (6.14) it follows that if $\varkappa_0$ equals zero at one point of $\{r\}$, then $C_0 = 0$, and at each other point of $\{r\}$ there is $\varkappa_0 = 0$. If at one point of $\{r\}$ there is $\varkappa_0 \neq 0$, then $C_0 \neq 0$, and at each other point of $\{r\}$ there is $\varkappa_0 \neq 0$. Because $\varkappa_0/U^2$ is the acceleration jump it follows that the discontinuity propagates along $\{r\}$. The curve $\{r\}$ is the acoustical ray (compare with optics). In general the acoustical ray is not orthogonal to s.

Denote by R^α the unit vector tangent to $\{r\}$. Direction of this vector is given by (6.12), Fig. 12.

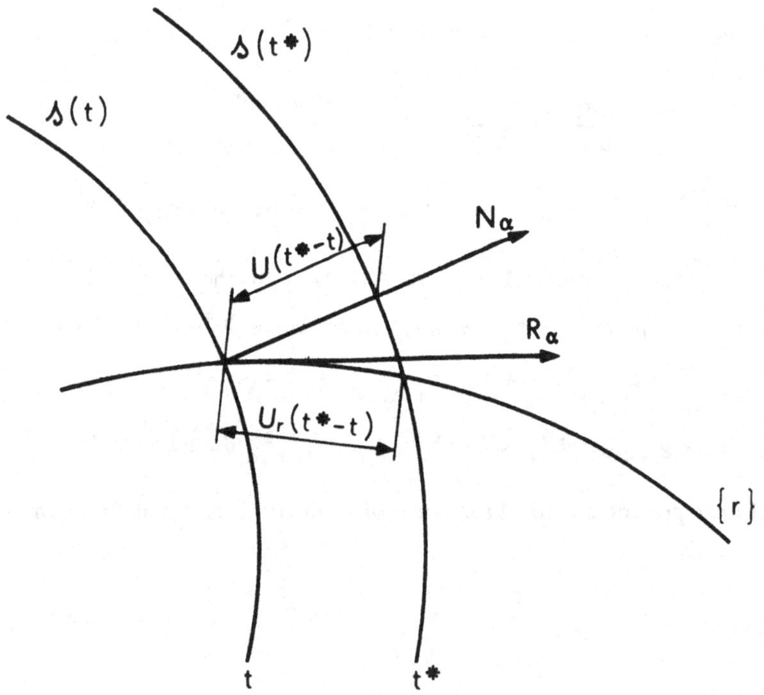

Fig. 12

Velocity of the surface Δ along the ray $\{r\}$ is the ray speed U_r. Considering two subsequent positions of (cf. Fig. 12) we have the approximate relation

$$t^* - t = \varphi\left(X^\alpha + U_r\left(t^* - t\right)R_\alpha\right) - \varphi\left(X^\alpha\right) =$$
$$= \varphi_{,\alpha} R^\alpha U_r\left(t^* - t\right).$$

Dividing by $\left(t^* - t\right)$ we arrive at the formula

$$U_r = \frac{1}{R^\alpha \varphi_\alpha}. \tag{6.15}$$

Because of the identity (cf. (2.3) and (2.7))

$$U = \frac{1}{N^\alpha \varphi_\alpha}$$

there is

$$U_r \geqslant U .$$

Pass to the calculation of $g_1^{\,i}$. Because $g_0^{\,i}$ has been already determined, the equation (6.8) is an algebraic equation with one unknown $g_1^{\,i}$. There holds the identity $\left(A_i{}^\alpha{}_k{}^\beta \Phi_\alpha \Phi_\beta - \varrho_R \delta_{ik}\right)A^k = 0$ and therefore $g_1^{\,i}$ can be obtained to within terms colinear with A^i. Basing on this remark represent $g_1^{\,i}$ in the following form

$$g_1^{\,i} = \varkappa_1 A^i + k_1^{\,i}, \qquad k_1^{\,i} \perp A^i . \tag{6.16}$$

The vector $k_1^{\,i}$ may be determined from (6.8). Assuming that this has been already done we look for the equation for $\varkappa_1$. Take (6.9) for $\nu = 0$, and multiply this equation by A^i The first term equals zero because of (3.9). Substituting (6.16) into the resulting equation we obtain the differential equation

$$\left(A_i{}^\alpha{}_k{}^\beta \Phi_\alpha \varkappa_{1,\beta} + A_i{}^\alpha{}_k{}^\beta \Phi_\beta \varkappa_{1,\alpha} + 2\varrho_R \delta_{ik} \varkappa_{1,t}\right) A^i A^k +$$
$$+ \varkappa_1 \left[A_i{}^\alpha{}_k{}^\beta\left(\Phi_\alpha A^k_{,\beta} + \Phi_\beta A^k_{,\alpha}\right) + 2\varrho_R \delta_{ik} A^k_{,t} + \left(A_i{}^\alpha{}_k{}^\beta \Phi_{\alpha,\beta} + A_i{}^\alpha{}_k{}^\beta{}_{,\alpha} \Phi_\beta\right)A^k\right]A^i =$$
$$= -\left[A_i{}^\alpha{}_k{}^\beta\left(\Phi_\alpha k^k_{1,\beta} + \Phi_\beta k^k_{1,\alpha}\right) + 2\varrho_R \delta_{ik} k^k_{1,t} +\right.$$
$$\left. + \left(A_i{}^\alpha{}_k{}^\beta \Phi_{\alpha,\beta} + A_i{}^\alpha{}_k{}^\beta{}_{,\alpha} \Phi_\beta\right) k_1^k\right] A^i - A^i \mathcal{L}_{ir}\, g_0^{\,r} .$$

The left-hand side of this equation is exactly the same, as the left-hand side of the equation (6.11), if $\varkappa_0$ is replaced by $\varkappa_1$. On the curve $\{b\}$ the right-hand side is a function of the parameter ν . Denoting this function by $K_1(\nu)$ we obtain the ordinary differential equation

(6.17)
$$\frac{d\varkappa_1}{d\nu} + \varkappa_1 P(\nu) = K_1(\nu)$$

where $P(\nu)$ is defined by (6.13). Solution of this equation consists of the special solution k_1 and the general solution of the corresponding homogeneous equation, therefore

$$\varkappa_1 = C_1 \exp\left(-\int_0^{\nu} P(\nu)\, d\nu\right) + k_1 .$$

Knowing $\varkappa_1$ we may take the equation (6.9) for $\nu = 1, 2, \ldots$ and determine g_2^k, g_3^k, ... Subsequent calculations lead to the following expressions for g_ν^k

$$g_\nu^k = \varkappa_\nu A^k + k_\nu^k \ , \qquad \nu = 1,2,3,\ldots$$

and to the equation for $\varkappa_\nu$

(6.19)
$$\frac{d\varkappa_\nu}{d\nu} + \varkappa_\nu P(\nu) = K_\nu(\nu) .$$

Finally the solution of (6.3) has the form

(6.20)
$$u^i = \sum_{\nu=0}^{\infty} S_\nu(\Phi)\left\{A^i C_\nu \exp\left(-\int_0^{\nu} P(\nu)\, d\nu\right) + k_\nu^i\right\} ,$$

where

(6.21)
$$k_\nu^i \perp A^i .$$

7. TRAVELLING WAVE

The calculations performed in the previous chapter were based on the decomposition (6.3). Essential was not the definition $(6.4)_1$ of $S_\nu(\Phi)$, but the property (6.5) of

$S_\nu(\Phi)$. It follows that if we take the set of arbitrary functions $T_\nu(\Phi)$, $\nu = 0, 1, 2, \ldots$ satisfying the reccursive formula

$$\frac{dT_\nu(\Phi)}{d\Phi} = T_{\nu-1}(\Phi). \tag{7.1}$$

we can immediately obtain other solution

$$u^i(X^\alpha, t) = \sum_{\nu=0}^{\infty} T_\nu(\Phi)\, g^i_\nu(X^\alpha, t), \tag{7.2}$$

where $g^i_\nu(X^\alpha, t)$ are the functions determined in the previous chapter. Take in particular

$$T_\nu(\Phi) = \frac{1}{(i\omega)^\nu}\, e^{i\omega\Phi} \tag{7.3}$$

where ω is arbitrary real parameter. It is evident that (7.3) satisfies (7.1). Substituting (7.3) into (7.2) we arrive at the complex solution of the linearized equations of motion (1.15)

$$u^i = \sum_{\nu=0}^{\infty} \frac{1}{(i\omega)^\nu}\, g^i_\nu\, e^{i\omega\Phi} \tag{7.4}$$

or

$$u^i = e^{i\omega(\Psi - t)} \sum_{\nu=0}^{\infty} \frac{1}{(i\omega)^\nu}\, g^i_\nu(X^\alpha, t).$$

In order to find the real solution note that by replacing ω by $-\omega$ we get solution adjoint to the solution (7.4). Adding both solutions we obtain the real solution

$$\begin{aligned} u^i(X^\alpha, t) &= \left(g^i_0 - \frac{1}{\omega^2}\, g^i_2 + \frac{1}{\omega^4}\, g^i_4 + \ldots\right)\cos\omega(\Psi - t) + \\ &+ \left(\frac{1}{\omega}\, g^i_1 - \frac{1}{\omega^3}\, g^i_3 + \frac{1}{\omega^5}\, g^i_5 + \ldots\right)\sin\omega(\Psi - t). \end{aligned} \tag{7.5}$$

Taking into account the expression (6.19) we get finally

$$u^i = A^i\left(x_0 - \frac{1}{\omega^2}\, x_2 + \frac{1}{\omega^4}\, x_4 + \ldots\right)\cos\omega(\Psi - t) +$$

$$(7.6)\qquad +\left(\frac{1}{\omega}\varkappa_1-\frac{1}{\omega^3}\varkappa_3+\frac{1}{\omega^5}\varkappa_5+\ldots\right)\sin\omega\left(\varphi-t\right)+s^i\ ,$$

where

$$s^i=\left(-\frac{1}{\omega^2}k_2^i+\frac{1}{\omega^4}k_4^i+\ldots\right)\cos\omega\left(\varphi-t\right)+$$
$$+\left(\frac{1}{\omega}k_1^i-\frac{1}{\omega^3}k_3^i+\ldots\right)\sin\omega\left(\varphi-t\right).$$

The vector s is usually small, as compared with the first term of (7.6), especially for big ω. In the subsequent calculations we drop out s^i and confine ourselves to the first term of (7.6). Denoting

$$M=\varkappa_0-\frac{1}{\omega^2}\varkappa_2+\ldots\ ,$$

$$(7.7)\qquad N=\frac{1}{\omega}\varkappa_1-\frac{1}{\omega^3}\varkappa_3+\ldots\ ,$$

$$\alpha=\operatorname{arc\,tg}\left(N/M\right)\ ,$$

we have

$$(7.8)\qquad u^i=A^i\sqrt{M^2+N^2}\cos\left[\omega\left(\varphi-t\right)-\alpha\right].$$

Expression (7.8) gives the displacement for a sinusoidal wave. At each fixed point X^α displacement u^i is sinusoidal function of time t. The wave (7.8) is the travelling wave, parameter ω is the angular frequency. The expression

$$\omega\left(\varphi-t\right)-\alpha\ ,$$

is the phase of this wave, and

$$A^i\sqrt{M^2+N^2}\ ,$$

its amplitude. The surfaces $\mathscr{S}_p$ of constant phase

$$(7.9)\qquad \omega\left(\varphi-t\right)-\alpha=\text{const.}$$

in general do not coincide with the discontinuity surface $\mathcal{S}$. The speed of $\mathcal{S}_p$ along the acoustical ray $\{r\}$ is the phase speed U_p , Fig. 13. Consider two positions of the surface $\mathcal{S}_p$ corresponding to the instants t and $t^* = t + \Delta t$. In accord with (7.9) there holds the identity

$$\left(\omega \Psi_{,\beta} - \alpha_{,\beta}\right) U_p R^\beta \Delta t - \omega \Delta t = 0 \ ,$$

and in the limit $\Delta t \to 0$

$$\left(\omega \Psi_{,\beta} - \alpha_{,\beta}\right) R^\beta U_p = \omega \ .$$

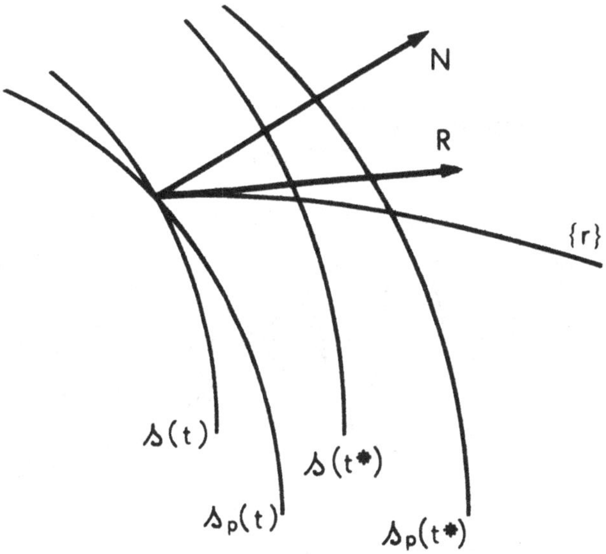

Fig. 13

Taking into account the relation (6.15) we have finally

$$U_p = U_r \left(1 - U_r R^\beta \frac{1}{\omega} \alpha_{,\beta}\right)^{-1}. \tag{7.10}$$

The expression (7.8) corrresponds to one, fixed frequency ω . Such a wave is monochromatic wave. The general wave consists of monochromatic waves of frequencies changing from ω_1 to $\omega_2 > \omega_1$. Consider superposition of two monochromatic waves with frequencies $\omega - \Delta\omega$ and $\omega + \Delta\omega$, $\Delta\omega \ll \omega$. In accord with (7.8) the resulting displacement is

$$v^i = u^i(X^\alpha, t, \omega + \Delta\omega) + u^i(X^\alpha, t, \omega - \Delta\omega),$$

or

$$v^i = 2A^i \sqrt{M^2 + N^2} \cos\left(\varphi - t - \frac{\partial \alpha}{\partial \omega}\right) \Delta\omega \cos\left[\omega(\varphi - t) - \alpha\right]. \tag{7.11}$$

This expression represents monochromatic wave of frequency ω with amplitude changing in space and time, Fig. 14.

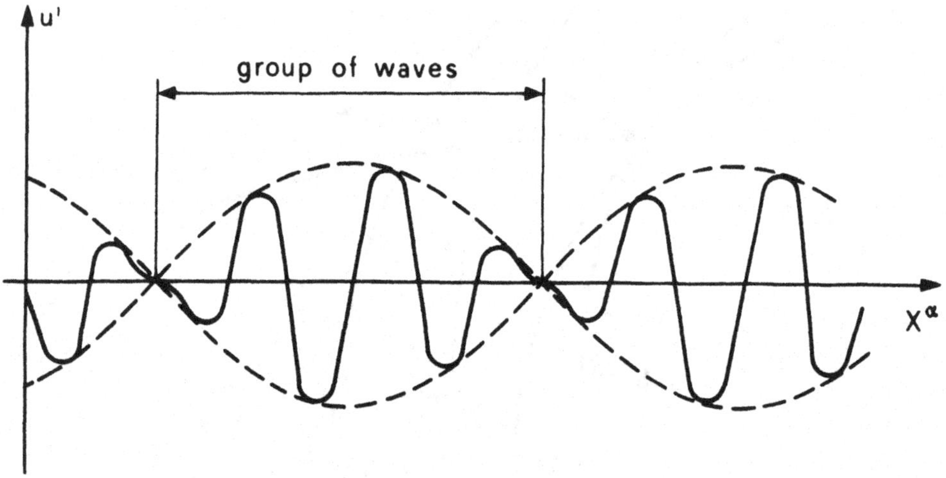

Fig. 14

The amplitude equals zero on the surface $\mathcal{S}_g$ given by the equation

$$\varphi - t - \frac{\partial \alpha}{\partial \omega} = 0 .$$

In general $\mathcal{S}$, $\mathcal{S}_p$ and $\mathcal{S}_g$ do not coincide. The speed of $\mathcal{S}_g$ along the acoustical ray $\{ r \}$ is the group speed U_g . In accord with the above equation there is

$$\left(\Phi_{,\beta} - \frac{\partial^2 \alpha}{\partial \omega \partial X^\beta}\right) R^\beta U_g \Delta t - \Delta t = 0 ,$$

and in the limit

$$\left(\Phi_{,\beta} - \frac{\partial^2 \alpha}{\partial \omega \partial X^\beta}\right) R^\beta U_g = 1 . \tag{7.12}$$

Taking into account the relation (6.15) we have finally

$$U_g = U_r \left[1 - U_r R^\beta \frac{\partial^2 \alpha}{\partial \omega \partial X^\beta} \right]^{-1} \tag{7.13}$$

Pass to the special case, when ω is sufficiently big to drop out all the terms $\varkappa_3 / \omega^3$, $\varkappa_4 / \omega^4$, ... In accord with (7.7) there hold the relations

$$M = \varkappa_0 - \frac{1}{\omega^2} \varkappa_2 , \quad N = \frac{1}{\omega} \varkappa_1 ,$$

$$\alpha = \operatorname{arc\,tg} \frac{\varkappa_1}{\omega \varkappa_0 - \varkappa_2/\omega} ,$$

$$\frac{\partial \alpha}{\partial X^\beta} = \frac{1}{\omega} \left(\frac{\varkappa_1}{\varkappa_0}\right)_{,\beta} , \quad \frac{\partial^2 \alpha}{\partial \omega \partial X^\beta} = - \frac{1}{\omega^2} \left(\frac{\varkappa_1}{\varkappa_0}\right)_{,\beta} .$$

The last two formulae were obtained after expanding arc tg into Taylor series. Substituting the above relations into (7.10) and (7.13) we arrive at the approximate formulae

$$U_p = U_r \left[1 - U_r R^\beta \frac{1}{\omega^2} \left(\frac{\varkappa_1}{\varkappa_0}\right)_{,\beta} \right]^{-1} ,$$

$$U_g = U_r \left[1 + U_r R^\beta \frac{1}{\omega^2} \left(\frac{\varkappa_1}{\varkappa_0}\right)_{,\beta} \right]^{-1} . \tag{7.14}$$

Expanding the fractions into Taylor series and droping out $1/\omega^3$, $1/\omega^4$ we obtain the identity

(7.15) $$U_p U_g = U_r^2$$

All the formulae given in the last chapters were derived, assuming that the coordinate systems are Cartesian. The formulae are valid in arbitrary coordinate systems if all the partial derivatives are replaced by total covariant derivatives or material time derivatives.

8. SPHERICAL WAVE IN ISOTROPIC MATERIAL

In order to illustrate the previous chapters consider the spherical wave in linear and isotropic elastic material. Assume that both $\{x^i\}$ and $\{X^\alpha\}$ are coinciding spherical coordinate systems $x^1 = X^1 = r$, $x^2 = X^2 = \vartheta$, $x^3 = X^3 = \varphi$ (no initial deformation). It is not more necessary to distinguish the greek and latin indices. Adopting the convention that all the indices are latin we have the following formulae for the metric tensor g_{ik} and the Christoffel symbols Γ_{ij}^{k}

(8.1) $$g_{ik} = \begin{bmatrix} 1 & 0 & 0 \\ 0 & r^2 & 0 \\ 0 & 0 & r^2\sin^2\vartheta \end{bmatrix} ,$$

$$\Gamma_{22}^{1} = -r \ , \quad \Gamma_{33}^{1} = -r\sin^2\vartheta \ , \quad \Gamma_{33}^{2} = -\sin\vartheta\cos\vartheta \ ,$$

$$\Gamma_{12}^{2} = \Gamma_{21}^{2} = \Gamma_{13}^{3} = \Gamma_{31}^{3} = \frac{1}{r} \ , \quad \Gamma_{23}^{3} = \Gamma_{32}^{3} = \mathrm{ctg}\,\vartheta \ .$$

The tensor $A_{i\ k}^{\ \alpha\ \beta}$ for isotropic elastic material has the form

(8.2) $$B_{i\ k}^{\ r\ s} = A_{i\ k}^{\ \alpha\ \beta}\delta_\alpha^r\delta_\beta^s = \lambda g_i^{\ r} g_k^{\ s} + \mu\left(g_{ik}g^{rs} + g_i^{\ s}g_r^{\ k}\right) .$$

where λ μ are Lame constants. Note that for homogeneous material $A_{i\ k}^{\ \alpha\ \beta}\|_\beta = 0$, $\varrho_R = \varrho =$ const., and therefore the equations of motion (1.15) reduce to

$$\mathscr{L}_{ir}\, u^r = B_i{}^{rs}_k\, u^k \big\|_{rs} - \varrho_R\, \ddot{u}_i = 0 \ . \tag{8.3}$$

Denoting $u_i = (u, v, w)$, $\partial u / \partial r = u_r$, ... we have the following equations of motion

$$\begin{aligned}
&(\lambda + 2\mu)\left(u_{rr} + \frac{2}{r} u_r - \frac{2}{r^2} u\right) + \\
&+ \frac{1}{r^2} \mu \left(u_{\vartheta\vartheta} + u_\vartheta \operatorname{ctg} \vartheta + \frac{1}{\sin^2\vartheta} u_{\varphi\varphi}\right) + \\
&+ \frac{1}{r^2} (\lambda + \mu) \left(v_{r\vartheta} + v_r \operatorname{ctg} \vartheta\right) - \frac{2}{r^3} (\lambda + 2\mu) \left(v_\vartheta + v \operatorname{ctg} \vartheta\right) + \\
&+ \frac{1}{r^2 \sin^2\vartheta} (\lambda + \mu) w_{r\varphi} - \frac{2}{r^3 \sin^2\vartheta} (\lambda + 2\mu) w_\varphi = \varrho \ddot{u} \ ,
\end{aligned} \tag{8.4}$$

$$\begin{aligned}
&(\lambda + 2\mu)\, u_{r\vartheta} + \frac{2}{r} (\lambda + 2\mu)\, u_\vartheta + \mu v_{rr} + \\
&+ \frac{\mu}{r^2 \sin^2\vartheta} v_{\varphi\varphi} + \frac{1}{r^2} (\lambda + 2\mu) v_{\vartheta\vartheta} + \frac{1}{r^2} (\lambda + 2\mu) v_\vartheta \operatorname{ctg} \vartheta + \\
&- \frac{1}{r^2 \sin^2\vartheta} (\lambda + 2\mu) v + \frac{1}{r^2 \sin^2\vartheta} (\lambda + \mu)\, w_{\vartheta\varphi} + \\
&- \frac{2}{r^2 \sin^2\vartheta} (\lambda + 2\mu)\, w_\varphi \operatorname{ctg} \vartheta = \varrho \ddot{v} \ ,
\end{aligned} \tag{8.5}$$

$$\begin{aligned}
&(\lambda + \mu) u_{r\varphi} + \frac{2}{r} (\lambda + 2\mu) u_\varphi + \\
&+ \frac{1}{r^2} (\lambda + \mu) v_{\vartheta\varphi} + \frac{1}{r^2} (\lambda + 3\mu) v_\varphi \operatorname{ctg} \vartheta + \mu w_{rr} + \\
&+ \frac{1}{r^2} \mu\, w_{\vartheta\vartheta} - \frac{1}{r^2} \mu w_\vartheta \operatorname{ctg} \vartheta + \frac{1}{r^2 \sin^2\vartheta} (\lambda + 2\mu)\, w_{\varphi\varphi} = \varrho \ddot{w} \ .
\end{aligned} \tag{8.6}$$

This system of equations is the basis for discussion of results. It is not used for consideration of the spherical wave.

Because the wave front is a sphere, there is

$$N_i = (1, 0, 0) \tag{8.7}$$

From (8.2) it follows that the acoustic tensor Q_{ik} and the propagation condition are (compare (3.7) and (3.8))

$$Q_{ik} = \begin{bmatrix} \lambda+2\mu & 0 & 0 \\ 0 & \mu r^2 & 0 \\ 0 & 0 & \mu r^2 \sin^2\vartheta \end{bmatrix}, \tag{8.8}$$

$$\left(\begin{bmatrix} \lambda+2\mu & 0 & 0 \\ 0 & \mu r^2 & 0 \\ 0 & 0 & \mu r^2 \sin^2\vartheta \end{bmatrix} - \varrho U^2 \begin{bmatrix} 1 & 0 & 0 \\ 0 & r^2 & 0 \\ 0 & 0 & r^2 \sin^2\vartheta \end{bmatrix} \right) \begin{bmatrix} A^1 \\ A^2 \\ A^3 \end{bmatrix} = 0 .$$

It follows from $(8.8)_2$ that there are three possible amplitudes, and three corresponding speeds

$$\begin{aligned} \overset{1}{A}{}^k &= (1, 0, 0), & \overset{1}{U}{}^2 &= (\lambda + 2\mu)/\varrho, \\ \overset{2}{A}{}^k &= (0, 1, 0), & \overset{2}{U}{}^2 &= \mu/\varrho, \\ \overset{3}{A}{}^k &= (0, 0, 1), & \overset{3}{U}{}^2 &= \mu/\varrho . \end{aligned} \tag{8.9}$$

First amplitude corresponds to the longitudinal wave. Both remaining waves are transverse. Take into account the longitudinal wave only. In order to simplify the notation we drop out the labelling index "1". For the wave considered we have

$$A^k = (1, 0, 0),$$
$$U^2 = (\lambda + 2\mu)/\varrho \tag{8.10}$$
$$\Psi = \frac{r}{U}, \quad \Phi = \frac{r}{U} - t.$$

Decompose the displacement u^i into the series (6.3), namely

$$u^i = \sum_{\nu=0}^{\infty} S_{\nu+2}\left(\frac{r}{U} - t\right) g^i_\nu (r, \vartheta, \varphi, t). \tag{8.11}$$

Substituting (8.10) into the equations of motion (1.15) we arrive at the system of equations (compare (6.7) – (6.9))

$$\left(\frac{1}{U^2} B_i{}^1{}_k{}^1 - \varrho g_{ik}\right) g_0^k = 0, \tag{8.12}$$

$$\left(\frac{1}{U^2} B_i{}^1{}_k{}^1 - \varrho g_{ik}\right) g_1^k +$$
$$+ \left(\frac{1}{U} B_i{}^1{}_k{}^s g_0^k\Big|_s + \frac{1}{U} B_i{}^s{}_k{}^1 g_0^k\Big|_s + 2\varrho\,\delta_{ik}\, g^k_{0,t}\right) = 0, \tag{8.13}$$

$$\left(\frac{1}{U^2} B_i{}^1{}_k{}^1 - \varrho g_{ik}\right) g^k_{\nu+2} + \left(\frac{1}{U} B_i{}^1{}_k{}^s g^k_{\nu+1}\Big|_s + \right.$$
$$\left. + \frac{1}{U} B_i{}^s{}_k{}^1 g_{\nu+1}\Big|_s + 2\varrho\,\delta_{ik}\, g^k_{\nu+1,t}\right) + \mathscr{L}_{ir}\, g^r_\nu = 0. \tag{8.14}$$

From (8.12) it follows

$$g_0^k = \varkappa_0 A^k, \tag{8.15}$$

where $\varkappa_0$ is a scalar parameter. Multiplying (8.13) by A_i we arrive at the differential equation

$$\left[\left(B_i{}^1{}_k{}^s + B_i{}^s{}_k{}^1\right)\varkappa_{0,s} + 2\varrho\delta_{ik}U\varkappa_{0,t}\right]A^iA^k +$$

(8.16)
$$+\varkappa_0 A^i\left[\left(B_i{}^1{}_k{}^s + B_i{}^s{}_k{}^1\right)A^k\Big|_s + 2\varrho\delta_{ik}UA^k_{,t}\right] = 0 .$$

Introduce the curve $\{b\}$ defined by the relations

$$\frac{dx^s}{d\nu} = \left(B_i{}^r{}_k{}^s + B_i{}^s{}_k{}^r\right)\Phi_r A^iA^k ,$$

$$\frac{dt}{d\nu} = 2\varrho g_{ik}A^iA^k .$$

In accord with (8.2) and (8.10) we have the relations

$$\frac{dr}{d\nu} = 2\varrho U , \quad \frac{d\vartheta}{d\nu} = \frac{d\varphi}{d\nu} = 0 ,$$

$$\frac{dt}{d\nu} = 2\varrho ,$$

which after integration lead to (Fig. 15)

(8.17)
$$r = 2\varrho U\nu , \quad \vartheta = \text{const.} , \quad \varphi = \text{const.} ,$$
$$t = 2\varrho\nu .$$

The equation (8.16) reduces now to the ordinary differential equation

(8.18)
$$\frac{d\varkappa_0}{dr} + \frac{\varkappa_0}{r} = 0 .$$

Assuming that for $r \to \infty$ there is $\varkappa_0 = 0$, we have

(8.19)
$$\varkappa_0 = C_0\frac{1}{r} ,$$

where C_0 is a constant. Substituting now (8.19) into (8.5) we obtain

$$\left(\frac{1}{U^2}B_i{}^1{}_k{}^1 - \varrho g_{ik}\right)g_1^k = 0 .$$

Comparing this equation with (8.8) we have

$$g_1^k = \varkappa_1 A^k . \tag{8.20}$$

In order to determine $\varkappa_1$ multiply (8.6) for $\nu = 0$ by A^l. The resulting equation is equivalent to the ordinary differential equation

$$\frac{d\varkappa_1}{d\nu} + \frac{\varkappa_1}{\nu} = \frac{C_0}{4\varrho^2 U \nu^3} .$$

Changing the variables in accord with (8.17) we get finally

$$\frac{d\varkappa_1}{dr} + \frac{\varkappa_1}{r} = \frac{C_0 U}{r^3} .$$

The general solution to this equation is the function

$$\varkappa_1 = \frac{C_1}{r} - C_0 \frac{U}{r^2} , \tag{8.21}$$

where C_1 is arbitrary constant.

Further calculations will give equations for $\varkappa_2$, $\varkappa_3$, $\varkappa_4$, ... The function $\varkappa_\nu$ may be expressed by all the constants C_0, C_1, C_2, ..., $C_{\nu-1}$ that appeared already in the formula for $\varkappa_{\nu-1}$ and additional constant C_ν. Finally we have

$$u^i = C_0 B_0^i + C_1 B_1^i + C_2 B_2^i + \ldots$$

where C_0, C_1, C_2, ... are arbitrary constants. Each of the coefficients B_0^i, B_2^i, B_3^i, ... is a solution to (8.3). The second derivative of B_0^i, third of B_1^i, fourth of B_2^i, ... are discontinuous on $\mathcal{S}$. Because we are interested in the acceleration wave assume $C_1 = C_2 = C_3 = \ldots = 0$. In this case all the equations for $\varkappa_1$, $\varkappa_2$, $\varkappa_3$, ... are homogeneous, therefore $\varkappa_2 = \varkappa_3 = \varkappa_4 = \ldots = 0$. In accord with (8.19) and (8.21) we arrive at the following expression for u^i and $\varkappa_0$, $\varkappa_1$

$$u^i = C_0 A^i \left[\frac{1}{r} S_2 \left(\frac{r}{U} - t \right) - \frac{U}{r^2} S_3 \left(\frac{r}{U} - t \right) \right] ,$$

$$\varkappa_0 = \frac{C_0}{r} , \quad \varkappa_1 = - \frac{C_0 U}{r^2} . \tag{8.22}$$

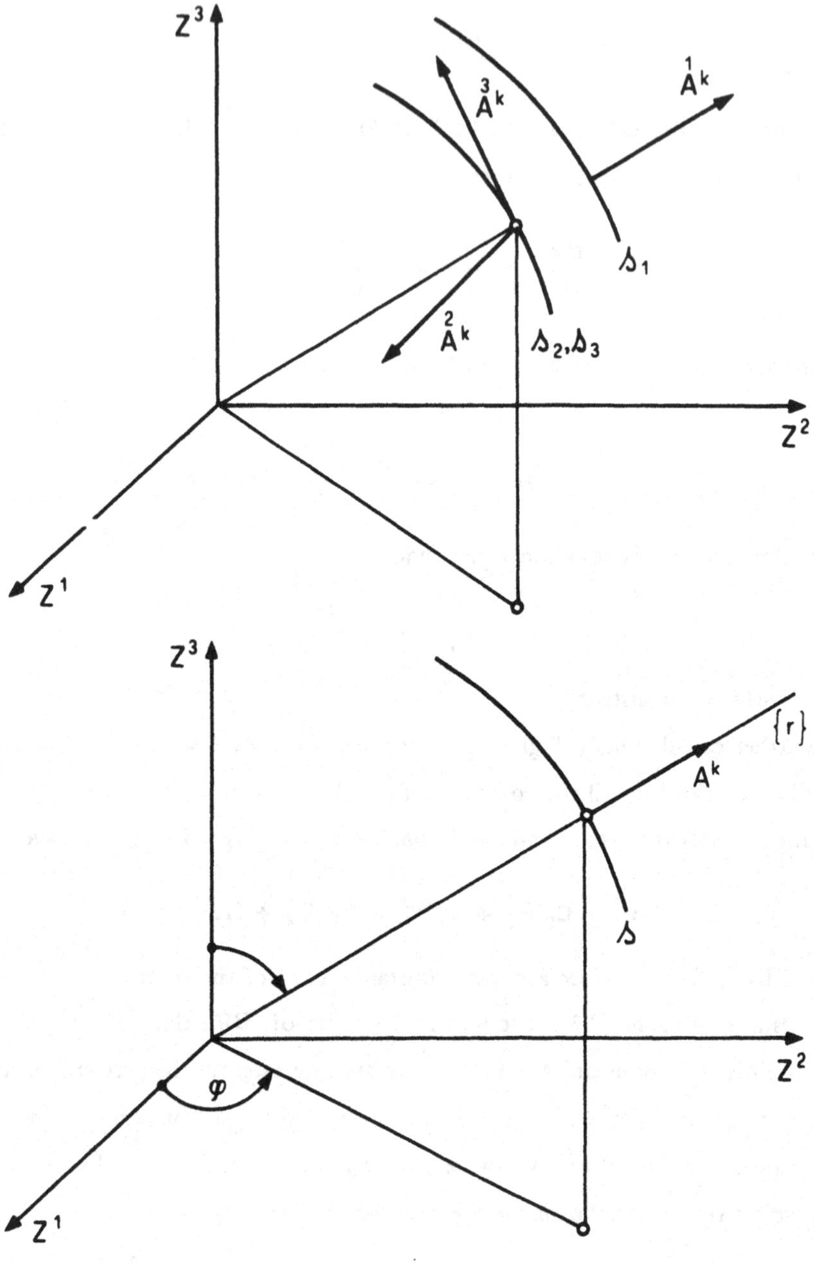

Z^3
$\overset{1}{A}{}^k$
$\overset{3}{A}{}^k$
$\overset{2}{A}{}^k$
$\mathcal{S}_1$
$\mathcal{S}_2, \mathcal{S}_3$
Z^2
Z^1
Z^3
$\{r\}$
A^k
$\mathcal{S}$
φ
Z^2
Z^1

Fig. 15

9. SPHERICAL PROGRESSIVE WAVE

Basing on the solution obtained in the previous chapter we construct the progressive wave

$$u = u^3 = C_0 \left(\frac{1}{r} - \frac{U}{i\omega r^2} \right) \exp \left[i\omega \left(\frac{r}{U} - t \right) \right], \quad u^1 = u^2 = 0 \, . \tag{9.1}$$

where ω is arbitrary parameter. It is easy to check, that (9.1) satisfies identically the equations of motion (8.5) - (8.6). There are many other solutions to this equations. The analysis given in Chapter 8 allowed to select one special solution (9.1) adjoined to the longitudinal spherical acceleration wave. It should be stressed, that the demand $u^2 = u^3 = 0$, $u^1 = u^1(r, t)$ does not lead to the solution (9.1). If we take for instance in the previous chapter $C_1 \neq 0$ we obtain solution of this type, but different from (9.1). There exist also other solutions, e.g. the solution

$$u^1 = \frac{1}{r} \mathcal{J}_{3/2} \left(\frac{\omega r}{U} \right) e^{i\omega t}, \quad u^2 = u^3 = 0 \, ,$$

representing standing wave.

The formula (9.1) gives the complex solution of the equations of motion. In order to obtain the real solution note that

$$u^* = C_0^* \left(\frac{1}{r} + \frac{U}{i\omega r^2} \right) \exp \left[-i\omega \left(\frac{r}{U} - t \right) \right], \quad u^2 = u^3 = 0 \, . \tag{9.2}$$

is also a solution (asterisk denotes complex conjugate). Adding together both above solutions (9.1) and (9.2) we arrive to the real solution

$$\tilde{u} = \tilde{u}^3 = \frac{1}{2} (u - u^*), \quad \tilde{u}^2 = \tilde{u}^3 = 0 \, ,$$

or after rearranging the terms

$$\tilde{u} = \frac{D_0}{r} \sqrt{1 - \frac{U^2}{\omega^2 r^2}} \cos \left[\omega \left(\frac{r}{U} - t \right) - a \right], \tag{9.3}$$

$$\alpha = \mathrm{arc\ tg}\left(-U/\omega r\right),$$

where D_0 is real constant. This expression represents sinusoidal wave. The phase is constant on the spheres on which

(9.4) $$\omega\left(\frac{r}{U} - t\right) - \alpha = \mathrm{const}.$$

This equation defines a function $r = r(t)$ determining the position of the surface Δ_p. Because ϑ = const., φ = const. is the acoustic ray, the derivative dr/dt is the phase speed U_p. Differentiating (9.4) with respect to time we obtain

(9.5) $$U_p = U\left[1 - \frac{U^2}{\omega^2 r^2 + U^2}\right]^{-1}.$$

It is evident that $U_p \geq U$. If $r \to \infty$ then $U_p \to U$.

Adding together two solutions corresponding to $\omega + \Delta\omega$ and $\omega\ \ \Delta\omega$ we obtain

$$u^1 = \frac{D_0}{r}\sqrt{1 - \frac{U^2}{\omega^2 r^2}}\cos\Delta\omega\left[\left(\frac{r}{U} - t\right) - \frac{rU}{\omega^2 r^2 + U^2}\right]\times$$

$$\times\cos\left[\omega\left(\frac{r}{U} - t\right) + \mathrm{arctg}\frac{U}{\omega r}\right],\quad u^2 = u^3 = 0.$$

This solution represents groups of waves (9.3) with the amplitude

$$\frac{D_0}{r}\sqrt{1 - \frac{U^2}{\omega^2 r^2}}\cos\Delta\omega\left[\left(\frac{r}{U} - t\right) - \frac{rU}{\omega^2 r^2 + U^2}\right].$$

The amplitude changes in space and time. It equals zero on the spheres Δ_g defined by the formula

(9.6) $$\frac{r}{U} - t - \frac{rU}{\omega^2 r^2 + U^2} = 0,$$

The speed of $\mathcal{S}_g$ along the acoustical ray ϑ = const., φ = const. is the group speed U_g. Differentiating (8.6) with respect to time we get

$$U_g = U\left[1 - \frac{U^2}{\omega^2 r^2 + U^2} + \frac{2\omega^2 r^2 U^2}{(\omega^2 r^2 + U^2)^2}\right]^{-1}. \tag{9.7}$$

10. VELOCITY WAVE

In the previous chapters there were considered the waves, for which the second, or higher derivatives were discontinuous on the singular surface. In elastic medium such waves are weak discontinuity waves, because such medium is governed by second order equation.

Assume that $\mathcal{S}_V$ is the front of the velocity wave. The equations of $\mathcal{S}_V$ are (for convenience we drop out the suscript V), cf. (2.1), (2.2)

$$X^\alpha = X^\alpha(M^K, t), \tag{10.1}$$

$$t = \varphi(X^\alpha). \tag{10.2}$$

The propagation velocity U_V of $\mathcal{S}_V$ is defined by the formula (cf. (2.7))

$$U_V = \left[\varphi_{,\varrho}\, \varphi_{,}{}^{\varrho}\right]^{-1/2}. \tag{10.3}$$

Because by assumption $x^i(X^\alpha, t)$ is continuous on in accord with (2.13) and (2.14) there is

$$\left[\!\left[x^i_{,\alpha} \right]\!\right] = H^i N_\alpha$$

$$\left[\!\left[\dot{x}^i \right]\!\right] = - H^i U_V. \tag{10.4}$$

The vector H^i will be called the amplitude of the velocity wave.

Pass to the equations of conservation of momentum and energy. Consider two positions of $\mathcal{S}_V$, corresponding to the instants t_1 and $t_2 > t_1$, and construct a cylinder between $\mathcal{S}_V(t_1)$ and $\mathcal{S}_V(t_2)$, Fig. 16.

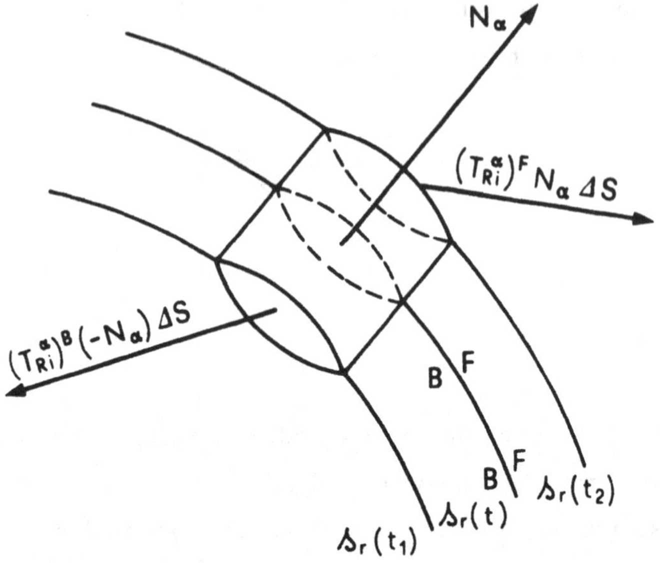

Fig. 16

For $t = t_1$ the momentum of the cylinder considered equals to (the whole cylinder is situated at the front side of $\mathcal{S}_V$)

$$\varrho_R U_V (t_2 - t_1) \Delta S \, (\dot{x}^i)^F$$

and for $t = t_2$ the momentum equals to (the cylinder is situated at the back side of $\mathcal{S}_V$)

$$\varrho_R U_V (t_2 - t_1) \Delta S \, (\dot{x}^i)^B$$

where ΔS denotes the area of the basis of the cylinder. It follows that the total increase of momentum in time $t_2 - t_1$ equals to

$$\varrho_R U_V (t_2 - t_1) \Delta S \left[\!\left[\dot{x}^i \right]\!\right] .$$

On the front surface of the cylinder there acts the force

$$(T_{Ri}^{\alpha})^F N_\alpha \Delta S \; ,$$

and on the back surface the force

$$(T_{Ri}^{\alpha})^B N_\alpha \Delta S \; .$$

For small $t_2 - t_1$ the forces acting on the side surface are negligible. Therefore the total force acting on the cylinder is

$$-[\![T_{Ri}^{\alpha}]\!] N_\alpha \Delta S \ .$$

The impulse of this force equals the increase of momentum, therefore

$$\varrho_R U_V (t_2 - t_1) \Delta S [\![\dot{x}^i]\!] = -[\![T_{Ri}^{\alpha}]\!] N_\alpha \Delta S (t_2 - t_1) \ .$$

Passing to the limit $\Delta S \rightarrow 0$, $t_2 \rightarrow t_1$ we obtain the equation of conservation of momentum

$$[\![T_{Ri}^{\alpha}]\!] N_\alpha = -\varrho_R U_V [\![\dot{x}_i]\!] \tag{10.6}$$

Consider in turn the conservation of energy. There holds the formula

$$\delta K + \delta \Sigma = \delta W + \delta Q \ , \tag{10.7}$$

where δK denotes increase of kinetic energy, δ increase of elastic energy, δW mechanical work and δQ the heat gained by the cylinder. Simple calculations lead to the formulae

$$\delta K = \frac{1}{2} \varrho_R U_V [\![\dot{x}_i \dot{x}^i]\!] \Delta S (t_2 - t_1) \ ,$$

$$\delta \Sigma = \varrho_R U_V [\![\sigma]\!] \Delta S (t_2 - t_1) \ ,$$

$$\delta W = -[\![T_{Ri}^{\alpha} \dot{x}^i]\!] N_\alpha \Delta S (t_2 - t_1) \ , \tag{10.8}$$

$$\delta Q = [\![Q^{\alpha}]\!] N_\alpha \Delta S (t_2 - t_1) \ .$$

The heat flux has been denoted by Q^α , and the specific elastic energy by σ . For elastic material $\sigma = \sigma(x^i_{\ \alpha}, \eta)$ where η is the entropy. Substituting (10.8) into (10.7) and passing to the limit $t_2 \rightarrow t_1$, $\Delta S \rightarrow 0$ we obtain finally

$$\frac{1}{2} \varrho_R U_V [\![\dot{x}_i \dot{x}^i]\!] + \varrho_R U_V [\![\sigma]\!] = -[\![T_{Ri}^{\alpha} \dot{x}^i]\!] N_\alpha - [\![Q^{\alpha}]\!] N_\alpha \ . \tag{10.9}$$

In accord with the second law of thermodynamics there holds the inequality

$$\varrho_R U_V \Delta S (t_2 - t_1)(\eta^B - \eta^F) \geq \left[\left(\frac{Q^\alpha}{T}\right)^B - \left(\frac{Q^\alpha}{T}\right)^F\right] N_\alpha \Delta S'(t_2 - t_1)$$

and in the limit $\Delta S \to 0, t_2 \to t_1$ the inequality

(10.10)
$$\varrho_R U_V [\![\eta]\!] \geq \left[\!\!\left[\frac{Q^\alpha}{T}\right]\!\!\right] N_\alpha .$$

The formulae derived above must be completed by the constitutive equations for elastic material, namely

(10.11)
$$\sigma = \sigma(x^i_\alpha, \eta),$$

(10.12)
$$T^\alpha_{Ri} = \varrho_R \frac{\partial \sigma}{\partial x^i_\alpha}, \quad T = \frac{\partial \sigma}{\partial \eta}.$$

They allow the express $[\![T^\alpha_{Ri}]\!]$ and $[\![\sigma]\!]$ by $[\![x^i_{\alpha}]\!]$ and $[\![\eta]\!]$. It is easy to see that the equations derived contain one redundant unknown. Therefore they allow to find one-parameter solution. In the next section we shall define a parameter possessing simple meaning, and find the solution for adiabatic wave.

11. ADIABATIC VELOCITY WAVE

The entropy inequality (10.10) leads to considerable difficulties when treating when treating the shock wave. Because all the interesting phenomena of velocity wave are already represented in the case $Q^\alpha = 0$, we confine ourselves to the adiabatic (but non isotropic) wave.

Assume that the wave is propagating into medium at rest, i.e.

(11.1)
$$(\dot{x}^i)^F = 0 .$$

From this assumption it follows

$$\left[\!\left[\dot{x}_i \dot{x}^i \right]\!\right] = \left[\!\left[\dot{x}_i \right]\!\right] \left[\!\left[\dot{x}^i \right]\!\right],$$
$$\left[\!\left[T_{Ri}^{\alpha} \dot{x}^i \right]\!\right] = \left[\!\left[T_{Ri}^{\alpha} \right]\!\right] \left[\!\left[\dot{x}^i \right]\!\right] + \left(T_{Ri}^{\alpha} \right)^F \left[\!\left[\dot{x}^i \right]\!\right]. \tag{11.2}$$

The function $\left(T_{Ri}{}^{\alpha} \right)^F$ represents initial stress and is known.

Repeat now all the essential equations from the previous chapter, taking into account (11.2) and the equality $Q^{\alpha} = 0$. We have

$$\left[\!\left[T_{Ri}^{\alpha} \right]\!\right]_{\alpha} = -\varrho_R U_V \left[\!\left[\dot{x}_i \right]\!\right], \tag{11.3}$$

$$\varrho_R U_V \left[\!\left[\sigma \right]\!\right] + \frac{1}{2} \varrho_R U_V \left[\!\left[\dot{x}^i \right]\!\right] \left[\!\left[\dot{x}^i \right]\!\right] =$$
$$= -\left[\!\left[T_{Ri}{}^{\alpha} \right]\!\right] \left[\!\left[\dot{x}^i \right]\!\right] N_{\alpha} - \left(T_{Ri}{}^{\alpha} \right)^F \left[\!\left[\dot{x}^i \right]\!\right] N_{\alpha}, \tag{11.4}$$

$$\left[\!\left[x^i{}_{\alpha} \right]\!\right] = H^i N_{\alpha}, \tag{11.5}$$

$$\left[\!\left[\dot{x}^i \right]\!\right] = -H^i U_V, \tag{11.6}$$

$$\sigma = \sigma\left(x^i{}_{\alpha}, \eta \right), \tag{11.7}$$

$$T_{Ri}{}^{\alpha} = \varrho_R \frac{\partial \sigma}{\partial x^i{}_{\alpha}}, \tag{11.8}$$

$$\left[\!\left[\eta \right]\!\right] \geqslant 0. \tag{11.9}$$

This is a set of 26 algebraic equations with 27 unknowns : $\left[\!\left[T_{Ri}{}^{\alpha} \right]\!\right]$, $\left[\!\left[x^i{}_{\alpha} \right]\!\right]$, $\left[\!\left[\dot{x}^i \right]\!\right]$, $\left[\!\left[\eta \right]\!\right]$, $\left[\!\left[\sigma \right]\!\right]$, H^i, and U_V. From all the solutions there must be selected the solutions satisfying (11.9), because only such solutions are physically admissible. It must be stressed that for the acceleration wave the equations (11.3) - (11.6) and the inequality (11.9) are satisfied identically.

Multiply (11.3) by $\left[\!\left[\dot{x}^i \right]\!\right]$ to get

$$-\frac{1}{2} \left[\!\left[T_{Ri}{}^{\alpha} \right]\!\right] \left[\!\left[\dot{x}^i \right]\!\right] N_{\alpha} = \frac{1}{2} \varrho_R U_V \left[\!\left[\dot{x}_i \right]\!\right] \left[\!\left[\dot{x}^i \right]\!\right],$$

and add this equation to (11.4). We obtain the equation

$$2\varrho_R U_V [\![\sigma]\!][\![T_{Ri}^{\alpha}]\!][\![\dot{x}^i]\!] N_\alpha = -2(T_{Ri}^{\alpha})^F [\![\dot{x}^i]\!] N_\alpha . \tag{11.10}$$

Starting from now on we shall use (11.10) instead of (11.4).

Pass to the calculation of the jumps of T_{Ri}^{α} and σ. There hold the formulae (cf. Fig. 17)

$$\begin{aligned} [\![T_{Ri}^{\alpha}]\!] = T_{Ri}^{\alpha}\ x^i{}_\alpha + [\![x^i{}_\alpha]\!],\ \eta + [\![\eta]\!] + \\ - T_{Ri}^{\alpha}(x^i{}_\alpha, \eta) . \end{aligned} \tag{11.11}$$

In order to obtain the propagation condition for the velocity wave assume that σ is analytic function of its arguments. From (11.8) it follows that T_{Ri}^{α} is analytic function, too. Denoting

$$\begin{aligned} \underbrace{\sigma_i{}^\alpha{}_k{}^\beta{}_m{}^\gamma{}^{\cdots}_{\cdots}}_{M} &= \left(\frac{\partial^M \sigma}{\partial x^i{}_\alpha\, \partial x^k{}_\beta\, \partial x^m{}_{\gamma \cdots}} \right)^F \\ \underbrace{\sigma_i{}^\alpha{}_k{}^\beta{}_m{}^\gamma{}^{\cdots}_{\cdots}}_{M}\, \underbrace{{}_{\eta \cdots \eta}}_{N} &= \left(\frac{\partial^{M+N} \sigma}{\partial x^i{}_\alpha\, \partial x^k{}_\beta\, \partial x^m{}_{\gamma \cdots}\, \partial \eta^N} \right)^F \end{aligned} \tag{11.12}$$

we have

$$\begin{aligned} [\![\sigma]\!] &= \sigma_i{}^\alpha [\![x^i{}_\alpha]\!] + \sigma_\eta [\![\eta]\!] + \\ &+ \frac{1}{2} \sigma_i{}^\alpha{}_k{}^\beta [\![x^i{}_\alpha]\!][\![x^k{}_\beta]\!] + \frac{1}{2} \sigma_i{}^\alpha{}_\eta [\![x^i{}_\alpha]\!][\![\eta]\!] + \ldots \end{aligned}$$

$$\begin{aligned} [\![T_{Ri}^{\alpha}]\!] &= \varrho_R\, \sigma_i{}^\alpha{}_k{}^\beta [\![x^k{}_\beta]\!] + \varrho_R\, \sigma_i{}^\alpha{}_\eta [\![\eta]\!] + \\ &+ \frac{1}{2} \varrho_R\, \sigma_i{}^\alpha{}_k{}^\beta{}_m{}^\gamma [\![x^k{}_\beta]\!][\![x^m{}_\gamma]\!] + \frac{1}{2} \varrho_R\, \sigma_i{}^\alpha{}_k{}^\beta{}_\eta [\![x^k{}_\beta]\!][\![\eta]\!] + \\ &+ \frac{1}{2} \sigma_i{}^\alpha{}_{\eta\eta} [\![\eta]\!]^2 + \frac{1}{6} \varrho_R\, \sigma_i{}^\alpha{}_k{}^\beta{}_m{}^\gamma{}_p{}^\delta [\![x^k{}_\beta]\!][\![x^m{}_\gamma]\!][\![x^p{}_\delta]\!] + \end{aligned}$$

$$+\frac{1}{6}\varrho_R\,\sigma_{i\ k\ m\ \eta}^{\ \alpha\ \beta\ \gamma}\left[\!\left[x^k_{\ \beta}\right]\!\right]\left[\!\left[x^m_{\ \gamma}\right]\!\right]\left[\!\left[\eta\right]\!\right]+$$

$$+\frac{1}{6}\varrho_R\,\sigma_{i\ k\ \eta\eta}^{\ \alpha\ \beta}\left[\!\left[x^k_{\ \beta}\right]\!\right]\left[\!\left[\eta\right]\!\right]^2+\frac{1}{6}\varrho_R\,\sigma_{i\ \eta\eta\eta}^{\ \alpha}\left[\!\left[\eta\right]\!\right]^3+\ldots$$

Substitute the above expressions into (11.3) and (11.10), denoting

$$\left[\!\left[\eta\right]\!\right]=S\ ,$$

and taking into account the relations (11.5) and (11.6). We arrive at the following two e-quations

$$\Big\{\sigma_{i\ k}^{\ \alpha\ \beta}H^kN_\beta+\sigma_{i\ \eta}^{\ \alpha}S+$$

$$+\frac{1}{2}\left[\sigma_{i\ k\ m}^{\ \alpha\ \beta\ \gamma}H^kH^mN_\beta N_\gamma+\sigma_{i\ k}^{\ \alpha\ \beta}H^kH_\beta S+\sigma_{i\ \eta\eta}^{\ \alpha}S^2\right]+$$

$$+\frac{1}{6}\left[\sigma_{i\ k\ m\ n}^{\ \alpha\ \beta\ \gamma\ \delta}H^kH^mH^nN_\beta N_\gamma N_\delta+\sigma_{i\ k\ m\ \eta}^{\ \alpha\ \beta\ \gamma}H^kH^mN_\beta N_\gamma S+\right.$$

$$\left.+\sigma_{i\ k\ \eta\eta}^{\ \alpha\ \beta}H^kN_\beta S^2-\sigma_{i\ \eta\eta\eta}^{\ \alpha}S^3\right]+$$

$$+\frac{1}{24}\left[\sigma_{i\ k\ m\ n\ p}^{\ \alpha\ \beta\ \gamma\ \delta\ \lambda}H^kH^mH^nH^pN_\beta N_\gamma N_\delta N_\lambda+\ldots\right]+\ldots\Big\}N_\alpha=U_V^2H_i\ ,$$

(11.13)

$$2\varrho_R U_V\Big\{\sigma_i^{\ \alpha}H^iN_\alpha+\sigma_\eta S+\frac{1}{2}\left[\sigma_{i\ k}^{\ \alpha\ \beta}H^iH^kN_\alpha N_\beta+\right.$$

$$\left.+\sigma_{i\ \eta}^{\ \alpha}H^iN_\alpha S+\sigma_{\eta\eta}S^2\right]+\frac{1}{6}\ \sigma_{i\ k\ m}^{\ \alpha\ \beta\ \gamma}H^iH^kH^mN_\alpha N_\beta N_\gamma+\ldots\Big\}+$$

$$+ \varrho_R U_V H^i N_\alpha \left\{ \sigma_i{}^\alpha{}_k{}^\beta H^k N_\beta + \sigma_i{}^\alpha{}_\eta S + \frac{1}{2} \left[\sigma_i{}^\alpha{}_k{}^\beta{}_m{}^\gamma H^k H^m N_\beta N_\gamma + \right. \right.$$

$$\left. \left. + \sigma_i{}^\alpha{}_k{}^\beta{}_\eta H^k N_\beta S + \sigma_i{}^\alpha{}_{\eta\eta} S^2 \right] + \frac{1}{6} \sigma_i{}^\alpha{}_k{}^\beta{}_m{}^\gamma{}_n{}^\delta H^k H^m H^n N_\beta N_\gamma N_\delta + \dots \right\} =$$

$$= 2 \left(T_{Ri}{}^\alpha \right)^F H^i N_\alpha U_V \; .$$

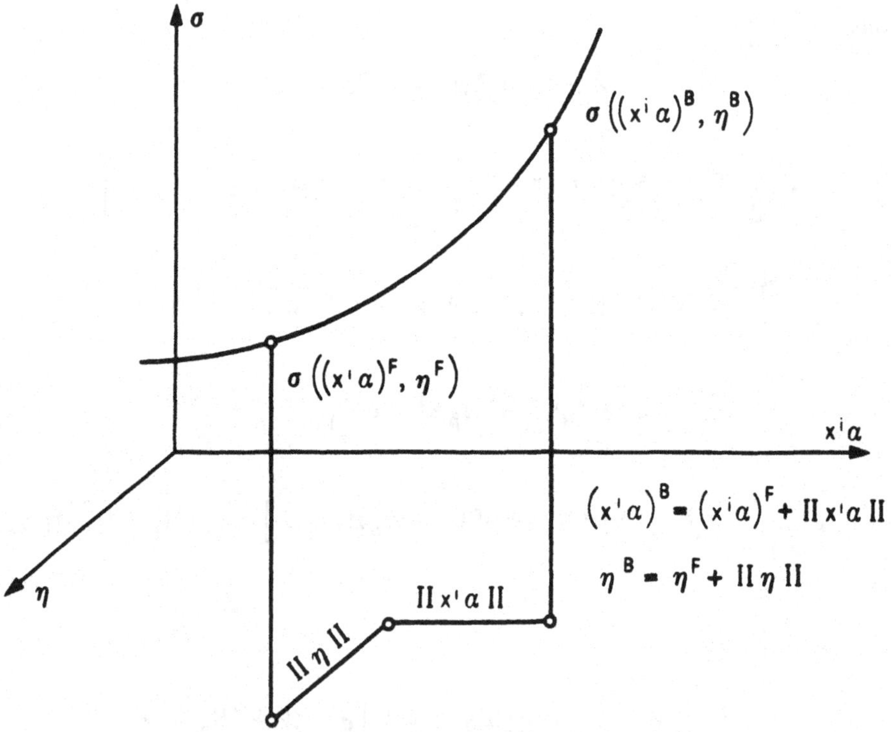

Fig. 17

The first and last term of the last equation cancel out because of (11.8). After rearranging the terms this equation has the following form

$$\frac{1}{6}\sigma_{i\;k\;m}^{\;\alpha\;\beta\;\gamma} H^i H^k H^m N_\alpha N_\beta N_\gamma + \frac{1}{12}\sigma_{i\;k\;m\;n}^{\;\alpha\;\beta\;\gamma\;\delta} H^i H^k H^m H^n N_\alpha N_\beta N_\gamma N_\delta +$$

$$+ S\left[2\sigma_\eta - \frac{1}{6}\sigma_{i\;k\;\eta}^{\;\alpha\;\beta} H^i H^k N_\alpha N_\beta + \ldots\right] +$$

$$+ S^2\left[\sigma_{\eta\eta} - \frac{1}{6}\sigma_{i\;\eta\eta}^{\;\alpha} H^i H_\alpha + \ldots\right] + \ldots = 0 \,. \tag{11.14}$$

The equations (11.13) and (11.14) constitute the set of four algebraic equations with five unknowns : H^i , S and U_V. Equation (11.14) allows to find the function $S = S(H_i)$. Because it is of infinite degree we shall look for the solution in the form of power series

$$S = C + C_i H^i + C_{ik} H^i H^k + \ldots , \tag{11.15}$$

where C , C_i , C_{ik} , . . . are constants. Note that if $H^i = 0$, then $[\![\dot{x}^i]\!] = [\![x^i_\alpha]\!] = 0$. Therefore $C = 0$. Substituting (11.15) into (11.14) and comparing the coefficient of the powers of H^i we obtain

$$C = C_i = C_{ik} = 0 \;,$$

$$C_{ikm} = \frac{1}{3}\sigma_{i\;k\;m}^{\;\alpha\;\beta\;\gamma} N_\alpha N_\beta N_\gamma / \sigma_\eta \;,$$

$$C_{ikmn} = \frac{1}{24}\sigma_{i\;k\;m\;n}^{\;\alpha\;\beta\;\gamma\;\delta} N_\alpha N_\beta N_\gamma N_\delta / \sigma_\eta \;,$$

and finally

$$S = \frac{1}{3}\sigma_{i\;k\;m}^{\;\alpha\;\beta\;\gamma} H^i H^k H^m N_\alpha N_\beta N_\gamma / \sigma_\eta +$$

$$+ \frac{1}{24}\sigma_{i\;k\;m\;n}^{\;\alpha\;\beta\;\gamma\;\delta} H^i H^k H^m H^n N_\alpha N_\beta N_\gamma N_\delta / \sigma_\eta + \ldots \tag{11.15}$$

Because of (11.9) there is

$$S \geqslant 0. \tag{11.16}$$

In accord with (11.15) the jump of entropy is of the order m^3 , where the parameter

$$m = \sqrt{H_i H^i} \tag{11.17}$$

is a measure of the intensity of the wave. Of course S^2 is of the order m^6 , and so on. Substituting (11.15) into (11.13) we arrive at the propagation condition of the velocity wave

$$\begin{aligned}
&\sigma_i{}^{\alpha}{}_k{}^{\beta} H^k N_\alpha N_\beta + \frac{1}{2} \sigma_i{}^{\alpha}{}_k{}^{\beta}{}_m{}^{\gamma} H^k H^m N_\alpha N_\beta N_\gamma + \\
&+ \frac{1}{6} \sigma_i{}^{\alpha}{}_k{}^{\beta}{}_m{}^{\gamma}{}_n{}^{\delta} H^k H^m H^n N_\alpha N_\beta N_\gamma N_\delta + \\
&+ \frac{1}{24} \sigma_i{}^{\alpha}{}_k{}^{\beta}{}_m{}^{\gamma}{}_n{}^{\delta}{}_p{}^{\lambda} H^k H^m H^n H^p N_\alpha N_\beta N_\gamma N_\delta N_\lambda + \\
&+ (1/\sigma_\eta)\left(\sigma_i{}^{\alpha}{}_\eta N_\alpha + \sigma_i{}^{\alpha}{}_k{}^{\beta} H^k N_\alpha N_\beta + \ldots\right)\left(\frac{1}{3} \sigma_m{}^{\gamma}{}_n{}^{\delta}{}_p{}^{\lambda} H^m H^n H^p N_\gamma N_\delta N_\lambda + \right. \\
&\left. + \frac{1}{4} \sigma_m{}^{\gamma}{}_n{}^{\delta}{}_p{}^{\lambda}{}_q{}^{\mu} H^m H^n H^p H^q N_\gamma N_\delta N_\lambda N_\mu + \ldots\right) = U_V^2 H_i \, .
\end{aligned} \tag{11.18}$$

It is seen that the direction of H^i and the velocity U_V are functions of m , $H^i = H^i(m)$, $U_V = U_V(m)$. We shall look for the solution of the equation (11.18) in the form of power series

$$\begin{aligned}
\frac{H^i(m)}{m} &= \overset{0}{H}{}^i + m \overset{1}{H}{}^i + m^2 \overset{2}{H}{}^i + \ldots \\
U_V(m) &= \overset{0}{U}_V + m \overset{1}{U}_V + m^2 \overset{2}{U}_V + \ldots
\end{aligned} \tag{11.19}$$

where $\overset{0}{H}{}^i$, $\overset{1}{H}{}^i$, . . . , $\overset{0}{U}_V$, $\overset{1}{U}_V$, . . . are constants. Substituting (11.19) into (11.18) we arrive at the infinite set of algebraic equations

$$\left(\sigma_i{}^\alpha{}_k{}^\beta N_\alpha N_\beta - \overset{0}{U}{}_V^2 g_{ik}\right)\overset{0}{H}{}^k = 0 \ , \tag{11.20}$$

$$\left(\sigma_i{}^\alpha{}_k{}^\beta N_\alpha N_\beta - \overset{0}{U}{}_V^2 g_{ik}\right)\overset{1}{H}{}^k - 2\,\overset{0}{U}_V \overset{0}{H}_i \overset{1}{U} + {}$$
$$+ \frac{1}{2}\,\sigma_i{}^\alpha{}_k{}^\beta{}_m{}^\gamma\, \overset{0}{H}{}^k \overset{0}{H}{}^m N_\alpha N_\beta N_\gamma = 0 \ , \tag{11.21}$$

$$\left(\sigma_i{}^\alpha{}_k{}^\beta N_\alpha N_\beta - \overset{0}{U}{}_V^2 g_{ik}\right)\overset{2}{H}{}^k - 2\,\overset{0}{U}_V \overset{0}{H}_i \overset{2}{U}_V - 2\,\overset{0}{U}_V \overset{1}{U}_V \overset{1}{H}_i - \overset{1}{U}{}_V^2 \overset{0}{H}_i + {}$$
$$+ \sigma_i{}^\alpha{}_k{}^\beta{}_m{}^\gamma\, \overset{0}{H}{}^k \overset{0}{H}{}^m N_\alpha N_\beta N_\gamma + \frac{1}{6}\,\sigma_i{}^\alpha{}_k{}^\beta{}_m{}^\gamma{}_n{}^\delta\, \overset{0}{H}{}^k \overset{0}{H}{}^m \overset{0}{H}{}^n N_\alpha N_\beta N_\gamma N_\delta - {}$$
$$+ \frac{1}{3}\,\sigma_i{}^\alpha{}_\eta N_\alpha\, \sigma_m{}^\gamma{}_n{}^\delta{}_p{}^\lambda\, \overset{0}{H}{}^m \overset{0}{H}{}^n \overset{0}{H}{}^p N_\gamma N_\delta N_\lambda \,/\, \sigma_\eta = 0 \ . \tag{11.22}$$

Because H^i/m is a unit vector there is

$$\overset{0}{H}_i \overset{0}{H}{}^i = 1 \ , \qquad \overset{0}{H}_i \overset{1}{H}{}^i = 0 \ , \qquad 2\,\overset{0}{H}_i \overset{2}{H}{}^i + \overset{1}{H}_i \overset{1}{H}{}^i = 0 \ . \tag{11.23}$$

Substituting (11.19) into the inequality (11.15) we obtain

$$\sigma_i{}^\alpha{}_k{}^\beta{}_m{}^\gamma\, \overset{0}{H}{}^i \overset{0}{H}{}^k \overset{0}{H}{}^m N_\alpha N_\beta N_\gamma \geq 0 \ . \tag{11.24}$$

In accord with (11.20) the vector $\overset{0}{H}{}^k$ is the proper vector of $\sigma_i{}^\alpha{}_k{}^\beta N_\alpha N_\beta$, and $\overset{0}{U}{}_V^2$ the proper value of $\sigma_i{}^\alpha{}_k{}^\beta N_\alpha N_\beta$.

In order to find $\overset{1}{U}_V$ multiply (11.21) by $\overset{0}{H}{}^i$. The first term drops out because of (11.20) and $\overset{1}{U}_V$ is determined by the equation

$$-2\,\overset{0}{U}_V \overset{1}{U}_V + \frac{1}{2}\,\sigma_i{}^\alpha{}_k{}^\beta{}_m{}^\gamma\, \overset{0}{H}{}^i \overset{0}{H}{}^k \overset{0}{H}{}^m N_\alpha N_\beta N_\gamma = 0 \ . \tag{11.25}$$

Taking into account the ineq .ality (11.24) we obtain

$$\overset{1}{U}_V = \frac{1}{4\overset{0}{U}_V}\,\sigma_{i\ k\ m}^{\ \alpha\ \beta\ \gamma}\,\overset{0}{H}^i\,\overset{0}{H}^k\,\overset{0}{H}^m\,N_\alpha N_\beta N_\gamma \geq 0 \ , \tag{11.26}$$

and finally, for small m , the formula (cf. (11.19))

$$U_V = \overset{0}{U}_V + \frac{m}{4\overset{0}{U}_V}\,\sigma_{i\ k\ m}^{\ \alpha\ \beta\ \gamma}\,\overset{0}{H}^i\,\overset{0}{H}^k\,\overset{0}{H}^m\,N_\alpha N_\beta N_\gamma \ .$$

For small intensities the propagation speed of the velocity wave increases with the intensity. In the limit $m \to 0$ the velocity wave propagates with the same speed, as the acceleration wave. In this case the amplitudes of the velocity wave and the acceleration wave are colinear (cf. (11.25)).

After substituting (11.26) inot (11.21) we arrive at the set of algebraic equations

$$\left(\sigma_{i\ k}^{\ \alpha\ \beta} N_\alpha N_\beta - \overset{0}{U}{}_V^2\, g_{ik}\right)\overset{1}{H}^k =$$

$$= \frac{1}{2}\,\sigma_{r\ k\ m}^{\ \alpha\ \beta\ \gamma}\,\overset{0}{H}^k\,\overset{0}{H}^m\left(\overset{0}{H}^r\,\overset{0}{H}_i - \delta_i^r\right) N_\alpha N_\beta N_\gamma \ .$$

Because the coefficient of $\overset{1}{H}^k$ is singular, represent $\overset{1}{H}^k$ in the form

$$\overset{1}{H}^k = a\,\overset{0}{H}^k + L^k \ , \qquad L^k \perp \overset{0}{H}^k \ ,$$

where a is arbitrary parameter, to be determined from (11.23). It is seen that $(11.23)_1$ leads to $a = 0$. After substituting $\overset{1}{H}^k = L^k$ and the expression (11.16) into (11.21) we can express L^k as a function of $\overset{0}{H}^i$. After multiplying (11.22) by $\overset{0}{H}^i$ the parameter $\overset{2}{U}$ may be determined. In this way we can find step-by-step all the coefficients of (11.19).

Consider now two acceleration waves moving at small distance in front of, and behind the velocity wave. The fronts of these waves denote by $\mathcal{S}^F$ and $\mathcal{S}^B$, respectively, Fig. 18. The propagation conditions of these waves (cf. (3.7)) are

$$\left[\left(A_{i\ k}^{\ \alpha\ \beta}\right)^F N_\alpha N_\beta - \varrho_R\left(U^F\right)^2 g_{ik}\right]\left(A^k\right)^F = 0 \ ,$$
$$\left[\left(A_{i\ k}^{\ \alpha\ \beta}\right)^B N_\alpha N_\beta - \varrho_R\left(U^B\right)^2 g_{ik}\right]\left(A^k\right)^B = 0 \ . \tag{11.27}$$

where $(A^k)^F$, $(A^k)^B$ are the amplitudes U^F , U^B the propagation speeds, and $\left(A_{i\ k}^{\ \alpha\ \beta}\right)^F$,

$\left(A_i{}^{\alpha}{}_k{}^{\beta}\right)^B$ the values of $A_i{}^{\alpha}{}_k{}^{\beta}$ in front, of and behind the velocity wave.

In accord with the notation (11.12) there is

$$\left(A_i{}^{\alpha}{}_k{}^{\beta}\right)^F = \varrho_R\, \sigma_i{}^{\alpha}{}_k{}^{\beta} .$$

Therefore the propagation condition (11.27) of $\mathcal{S}^F$ has the form

$$\left[\sigma_i{}^{\alpha}{}_k{}^{\beta} N_\alpha N_\beta - \left(U^F\right)^2 g_{ik}\right]\left(A^k\right)^F = 0 .$$

By comparing with (11.20) we infer that

$$U^F = \overset{0}{U}_V , \quad \left(A^k\right)^F = \overset{0}{H} . \tag{11.28}$$

Because $\overset{0}{U}_V$ is the speed of velocity wave for $m \to 0$ we have the interesting result : Infinitesimal velocity wave propagates with the speed of the acceleration wave.

Consider now the acceleration wave $\mathcal{S}^B$. Expanding $\left(A_i{}^{\alpha}{}_k{}^{\beta}\right)^B$ into Taylor series and making use of (11.5) we have

$$\left(A_i{}^{\alpha}{}_k{}^{\beta}\right)^B = \left(A_i{}^{\alpha}{}_k{}^{\beta}\right)^F + \left(\frac{\partial}{\partial x^m{}_\gamma} A_i{}^{\alpha}{}_k{}^{\beta}\right)^F \left[\!\left[x^m{}_\gamma \right]\!\right] =$$

$$= \varrho_R\, \sigma_i{}^{\alpha}{}_k{}^{\beta} + \varrho_R\, \sigma_i{}^{\alpha}{}_k{}^{\beta}{}_m{}^{\gamma} H^m N_\gamma .$$

Therefore the propagation condition $(11.27)_2$ has the form

$$\left[\sigma_i{}^{\alpha}{}_k{}^{\beta} N_\alpha N_\beta - \left(U^F\right)^2 g_{ik}\right]\left(A^k\right)^F = 0 . \tag{11.29}$$

It is evident that the solution $U^B, \left(A^k\right)^B$ of this propagation condition depends on m . In order to find this solution represent both functions in form of the power series

$$\left(A^k\right)^B = \left(\overset{0}{A}{}^k\right)^B + m\left(\overset{1}{A}{}^k\right)^B + \dots ,$$

$$U^B = \overset{0}{U}{}^B + m\,\overset{1}{U}{}^B + \dots ,$$

where the coefficients of m^n are constants. Substitute this series and $(11.19)_1$ into the propagation condition (11.29) to obtain

$$\Big[\sigma_i{}^{\alpha}{}_k{}^{\beta} + m\,\sigma_i{}^{\alpha}{}_k{}^{\beta}{}_m{}^{\gamma} N_\alpha N_\beta N_\gamma \overset{0}{H}{}^m + \dots$$

$$- \left(\overset{0}{U}{}^B + m\,\overset{1}{U}{}^B + \dots\right) g_{ik}\Big]\left[\left(\overset{0}{A}{}^k\right)^B + m\left(\overset{1}{A}{}^k\right)^B + \dots\right] = 0 .$$

Equating to zero all the coefficients of m^0, m, m^2, ... we arrive to infinite set of equations. The first two equations are the following

$$\left[\sigma_i{}^{\alpha}{}_k{}^{\beta} N_\alpha N_\beta - \left(\overset{0}{U}{}^B\right)^2 g_{ik}\right]\left(\overset{0}{A}{}^k\right)^B = 0 \ , \tag{11.30}$$

$$\left[\sigma_i{}^{\alpha}{}_k{}^{\beta}{}_m{}^{\gamma} N_\alpha N_\beta - \left(\overset{0}{U}{}^B\right)^2 g_{ik}\right]\left(\overset{1}{A}{}^k\right)^B +$$

$$- \sigma_i{}^{\alpha}{}_k{}^{\beta}{}_m{}^{\gamma} N_\alpha N_\beta N_\gamma \overset{0}{H}{}^m \left(\overset{0}{A}{}^k\right)^B - 2\overset{0}{U}{}^B \overset{1}{U}{}^B \left(\overset{0}{A}{}^k\right)^B g_{ik} = 0 \ . \tag{11.31}$$

Take into account (11.28) and compare (11.30) with (11.20). There is

$$\overset{0}{U}{}^B = \overset{0}{U}_V \ , \quad \left(\overset{0}{A}{}^k\right)^B = \overset{0}{H}{}^k \ . \tag{11.32}$$

Multiply (11.31) by $\overset{0}{H}{}^i$ and take into account (11.30), $(11.32)_2$. We obtain

$$\overset{1}{U}{}^B = \frac{1}{2\overset{0}{U}_V} \sigma_i{}^{\alpha}{}_k{}^{\beta}{}_m{}^{\gamma} \overset{0}{H}{}^i \overset{0}{H}{}^k \overset{0}{H}{}^m N_\alpha N_\beta N_\gamma \ , \tag{11.33}$$

and finally

$$U^B = \overset{0}{U}_V + \frac{m}{2\overset{0}{U}_V} \sigma_i{}^{\alpha}{}_k{}^{\beta}{}_m{}^{\gamma} \overset{0}{H}{}^i \overset{0}{H}{}^k \overset{0}{H}{}^m N_\alpha N_\beta N_\gamma \ . \tag{11.34}$$

Joining together (11.26), (11.28) and (11.34) we obtain the inequalities

$$U^B \geqslant U_V \geqslant U^F \ . \tag{11.35}$$

The quantities U^F and U^B are the sound speeds for the medium in front of and behind the velocity wave. The velocity wave is therefore supersonic with respect to the medium in front of it, and subsonic with respect to the medium behind it.

In the linear elasticity

$$U^F = U^B = \sqrt{\frac{\lambda + 2\mu}{\varrho}} \quad \left(\text{or } \sqrt{\frac{\mu}{\varrho}}\right) . \tag{11.36}$$

It follows from (11.35) that in this case the propagation speed of velocity wave is

$$U_V = \sqrt{\frac{\lambda + 2\mu}{\varrho}} \qquad \left(\text{or } \sqrt{\frac{\mu}{\varrho}}\right),$$

and is independent of the intensity of this wave.

The reader is expected to be familiar with tensor calculus. The detailed exposition of the theory of double point tensor fields is given in the paper by J.L. Ericksen in Handbuck der Physik III/1, Berlin 1060.

Finite elasticity is exposed in the monograph by C. Truesdell, R. Toupin in Handbuch der Physik III/1, Berlin 1960 and monograph by C. Truesdell, W. Noll in Handbuch der Physik III/3, Berlin 1965.

ADDITIONAL LITERATURE

1. Bland D.R., Nonlinear dynamic elasticity, Waltham 1969.

2. Fedorov I., Theory of elastic waves in cristals, New York 1968.

3. Brillouin L., Wave propagation and group velocity, New York 1960.

4. Courant R., Hilbert D., Methoden der matematischen Physik II, Berlin 1968.

5. Wesolowski Z., Linear independence of amplitudes of reflected acceleration waves, Bull. Acad. Polon. Sci., Ser, Sci. Tech., 12, 23 , 1975.

ON THE NON-LINEAR BEHAVIOUR AND THE STABILITY OF RETICULATED ELASTIC SYSTEMS

by

S. J. BRITVEC
Professor of Engineering Mechanics
University of Stuttgart
and
University of Zagreb

1. GENERAL PROPOSITIONS

1.1. Elastic Systems Composed of Finite Elements

A variety of physical structural elements may be imagined without difficulty. These elements, when jointed together by means of certain types of connections, form a structural system with definite elastomechanical properties. The geometrical shape of such a structural assembly may vary from a beam to a three-dimensional shell-type structure. For example, cubic or triangular elements may be imagined with, as yet, unspecified internal properties Fig.1.1 and 1.2 which may be assembled into arrays

---

* *This text is based on a series of lectures delivered at the International Centre for Mechanical Sciences (CISM) at Udine, Italy, from October [illegible].*

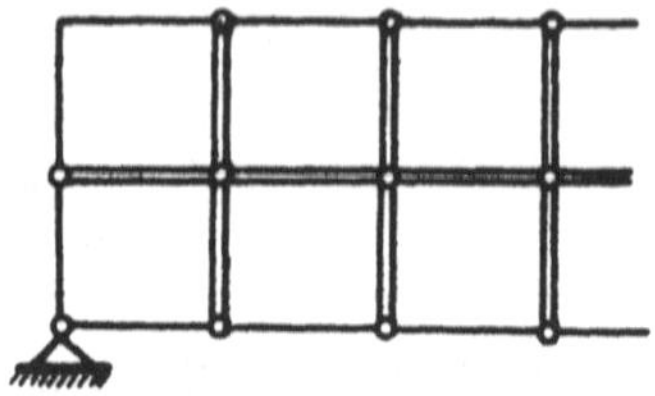

Fig. 1.1.

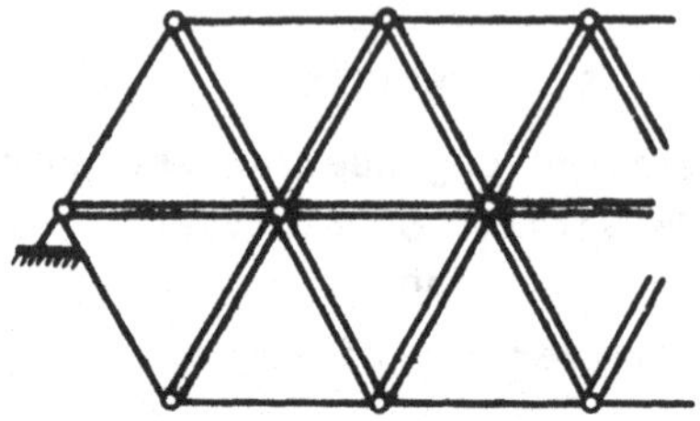

Fig. 1.2.

forming a reticulated structure. These arrays may be in a plane, such as those in Fig.1.1 and 1.2, or they may be in space, Fig.1.3. Space assemblies in the shape of reticulated shells are particularly useful in structural applications. For example, Fig.1.3 shows two "parallel" hyperbolic paraboloid surfaces generated by straight lines which enclose physical elements. The location of a typical element is indicated by the contours (a b c d) and (a'b'c'd'). Any such assembly of elements may be described as a structural system if appropriately supported so that external loads may be applied to it. For example, if the cubes are stiffened in such a way that the outside edges are formed of high-tensile steel bars, while

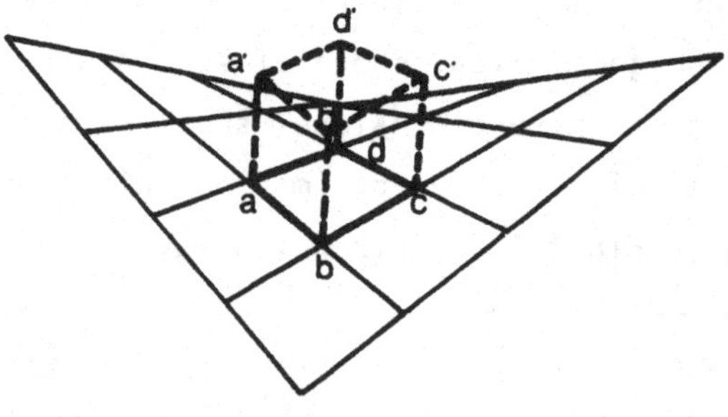

Fig. 1.3.

on the inside additional bars are introduced, then *braced elements* result. These elements may be formed in the shape of tetrahedrons, octahedrons, combinations of cubes and tetrahedrons etc., resulting in highly hyperstatic systems. A typical system of this kind is shown in Fig.1.4a in the form of a cooling tower made from a combination of octahedrons and tetrahedrons. A hyperbolic paraboloid reticulated shell, made from a combination of cube-tetrahedron elements is shown in Fig.1.4b.

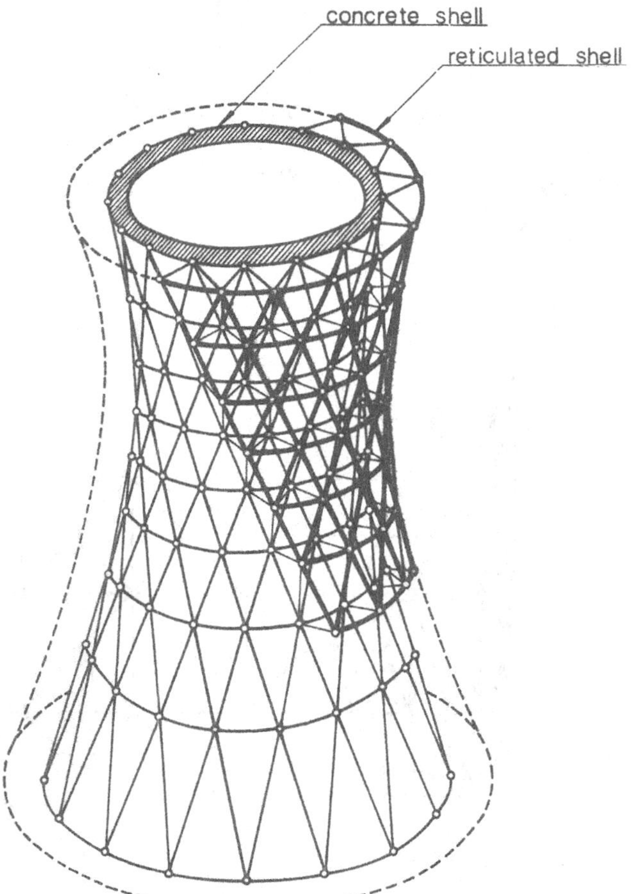

HYPERBOLIC - PARABOLOID SHELL

Fig. 1.4.a

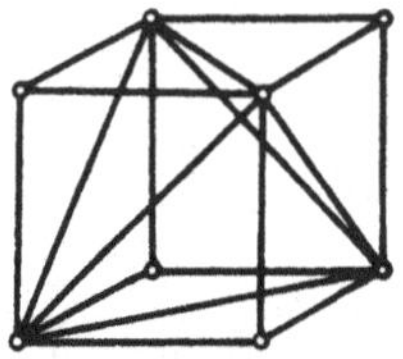

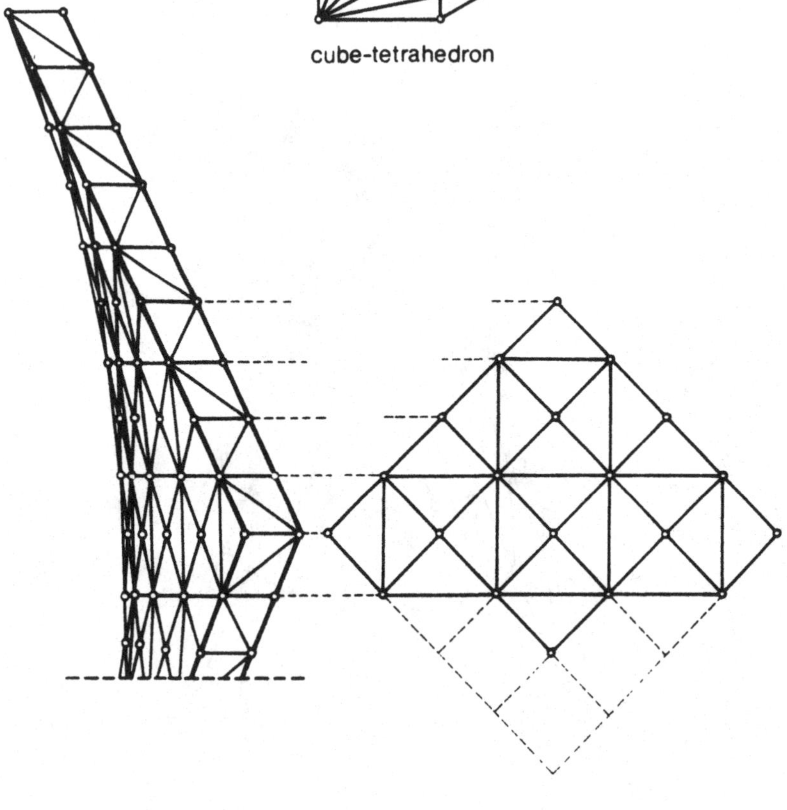

HYPERBOLIC PARABOLOID SHELL
(one layer of elements)
Fig.1.4.b.

2. SYMMETRICAL AND NON-SYMMETRICAL ELEMENTS - EXTERNAL LOADING

Generally, loads concentrated centrally at the joints between the elements need be considered. Moreover, we shall suppose these loads to be *conservative*. Actually, the loading may be distributed over an envelope (continuous shell or membrane) of the shell-type assembly and transmitted to the space lattice only through the points (joints) of contact between this envelope and the reticulated shell.

Elements made from practically *inextensible* slender members are of particular interest. With this assumption the analysis of stability is considerably simplified. There are further differences in the analytical treatment, *depending on the mechanical properties of the connections*. If these are *simple pins*, a particular class of the so-called *symmetrical systems* is derived (this symmetry is not optical, but is conceived in regard to the total potential energy changes about a certain state). If, on the other hand, the connections are rigid, the system is referred to as non-symmetrical.

The deformations of braced elements may be described by the end-rotations (in flexure and in torsion) of the slender members at the corners (joints) of the elements, which are then considered as the generalized coordinates of deformation of the assembled system in the sense of Lagrange.

In the case of symmetrical systems, the flexural deformations of inextensible pin-ended bars may be described adequately by their end-tangent rotations or lateral deflections, which are then sufficient for analysis of an entire symmetrical system.

Such structural assemblies may buckle *statically* in different modes in which certain *kinematic mechanisms* within the system are generated. Each mechanism depends, generally, non-linearly on certain geometrical quantities or *coordinates associated with the buckling mode*. Therefore, the states of deformation associated with these buckling modes may be *statically unstable* and this is an important consideration in analysing the statical or dynamic behaviour of reticulated shells.

To describe mathematically the mechanical behaviour of a system of this type we define, following the concepts of Lagrange, the coordinates of deformation of the system by

$$q_i \quad ; \; i = 1,2,\ldots, n$$

(Occasionally this symbol may be used to characterize a coordinate *change* referred to the initial value of this coordinate, then denoted by the capital letter Q_i or Q_i^o.) Then, in the case od a pin-ended prismatic bar, q_i represents (conveniently) the rotation of the end-tangents.

It should be pointed out that the flexural shortening e_i of an inextensible bar, could not, as a matter of course, be chosen to represent a generalized coordinate, since e_i is proportional to q_i^2 in this case. But we shall see later on, that, indeed, for some systems, referred to as the symmetrical systems, e_i can in fact be used as a new coordinate of deformation, while the relationship $e_i \propto q_i^2$ holds. Buckling of a typical simple plane symmetrical system is shown in Fig.2.1. It is evident from this case, that generally, the displacements of the joints of a symmetrical element or system are independent of the sign of the tangent rotations of the buckling members and so depend only on the even powers of these rotations or coordinates. Here, the corresponding displacement of the load is proportional to the flexural contraction of the buckling bar.

Consider next the deformations of some non-symmetrical elements or simple systems in a plane. A rigidly-jointed triangular element (1),(2), (3), consisting of two elastic and one rigid bar is shown in Fig.2.4. If a load is applied to the free joint (1), then the buckled form of this element may assume the shapes indicated in figures 2.4a and 2.4b. Here, the clokwise or the countercloskwise rotation of the joint may be considered as the *mode of buckling*.

It is to be noted that each of these two shapes may exist in statical equilibrium under a centrally applied load at joint (1), but that in each case this load is *different*. In fact, it can be shown (see Figs.2.2 and 2.3), that in order that statical equilibrium of shape (a) may be

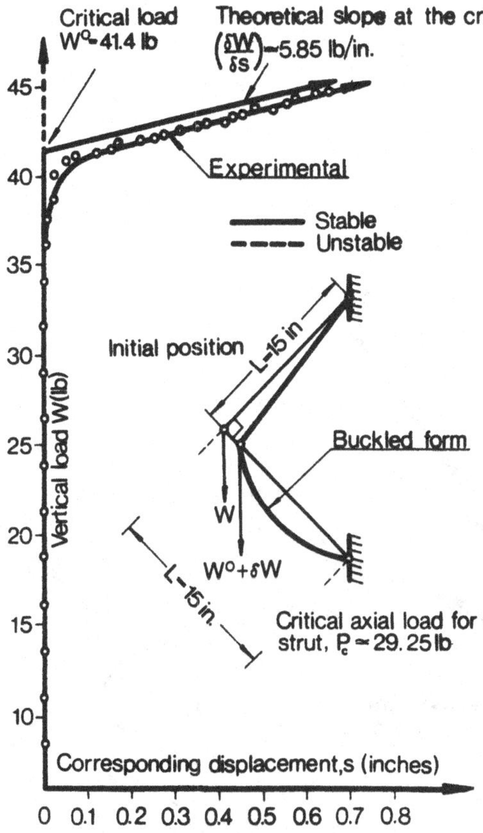

Fig. 2.1.**

maintained, always a higher load is required than that necessary to maintain in equilibrium the shape (b). Moreover, these two loads need not rotate for this equilibrium to exist, i.e. buckled shapes, such as these, *exist under conservative external loads applied centrally at the joints.*

This behaviour is in stark contrast to that of a similar symmetrical system typified in Fig. 2.1.

Let us point out that also in space the rotations of the joints of non-symmetrical systems represent natural or *physical*, sometimes called the non-orthogonal* coordinates of deformation.

* *For a detailed explanation see Ref. 12, op.cit.*

***Figs 2.1 to 2.3 reproduced by permission.* [12]

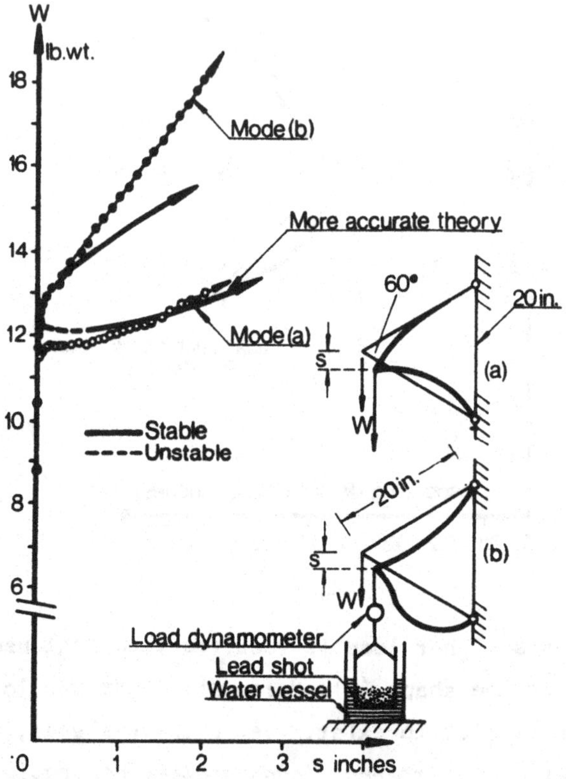
W
lb.wt.
18
16
14
12
10
8
6
Mode (b)
More accurate theory
Mode (a)
Stable
Unstable
60°
20in.
(a)
s
W
20in.
(b)
s
W
Load dynamometer
Lead shot
Water vessel
0
1
2
3
s inches

Fig 2.2

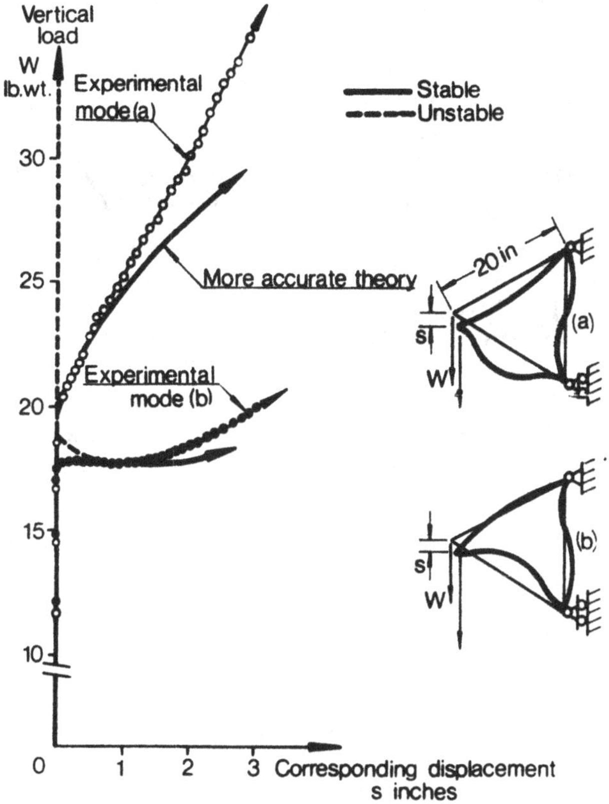
Vertical load
W lb.wt.
Experimental mode (a)
Stable
Unstable
30
25
20
15
10
More accurate theory
Experimental mode (b)
20 in
(a)
(b)
s
W
0
1
2
3
Corresponding displacement s inches

Fig. 2.3.

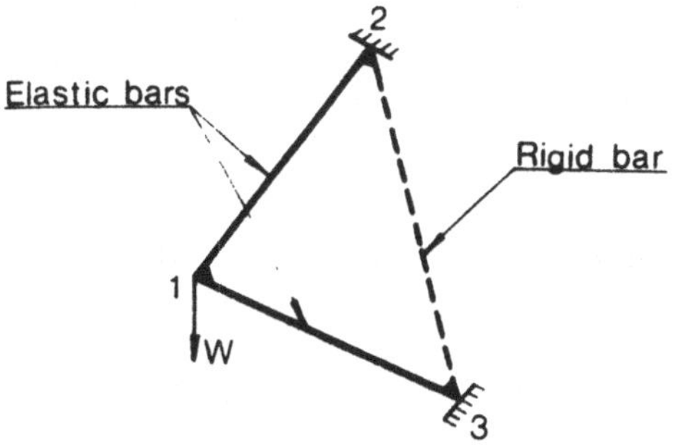
Elastic bars
Rigid bar
1
2
3
W

Fig. 2.4.

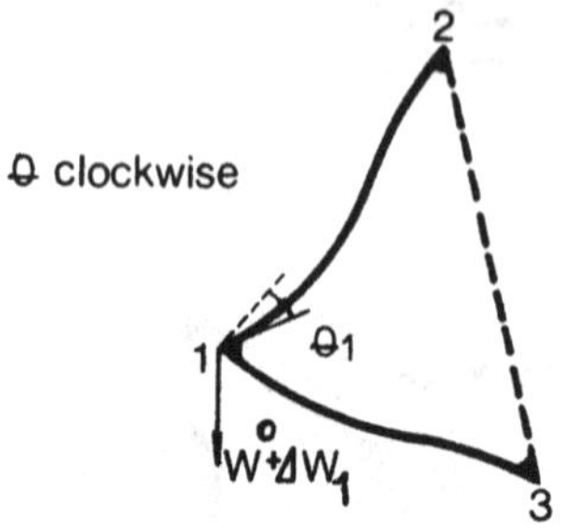

Fig. 2.4. (a)

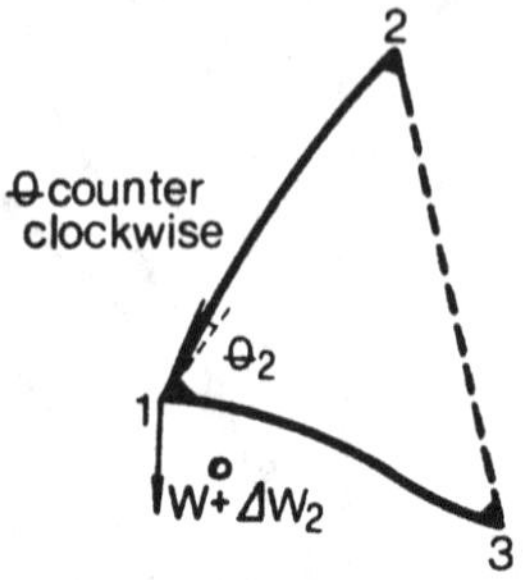

Fig. 2.4. (b)

3. THE GENERAL THEORY

3.1. The Equilibrium Paths of Non-symmetrical Systems in Physical Coordinates

In the following development we confine our attention to some of the simplest non-linear equilibrium paths of non-symmetrical and symmetrical systems. We have already defined non-symmetrical systems as being composed of elements with non-symmetrical properties of flexural deformation. We stipulate some changes in the coordinates of deformation of such a system, which we denote by the symbols q_i, $i = 1,2,\ldots,n$. This stipulation is quite general and, in fact, no physical visualization of these coordinates is necessary at the start to arrive at some general results. It is sufficient that the deformations of the individual elements and of the system may be described *completely* by *a finite* number of such coordinate changes referred to an initial state.

We further assume the existence of the total potential energy function which then depends on these coordinates and in some way on the external loading. It is to be noted here that a change in the total potential energy V is not equivalent to the change in the elastic energy of the system, but that the change of the *potential* energy of the external loads is *included* in this definition.

Consider now a conservative elastic system, such that its external loads can be derived from a potential function H. Let us denote the elastic energy of this system by the function U, where U now depends only on the coordinates of deformation Q_i and on the elastic constants of the material. A change in V in relation to the initial state is given by:

$$\Delta V = \Delta U - \Delta H \qquad (3.1)$$

It remains to define more precisely the external loads of this system. If we consider proportionate loading systems, such that all the external load-changes are proportional to a load-parameter change p, then also ΔH depends on p and on the coordinate changes q_i, so that we may write ΔV as a function

$$\Delta V = \Delta V\,(Q_i^o + q_i,\ P^o + p); \quad i = 1,2,\ldots,n. \tag{3.2}$$

where Q_i^o are the initial values of the generalized coordinates of the system. Here, the only variables are the *changes* q_i and p.

The next object of our analysis is to establish an expression for the relation (3.2) so that it can be used to study the equilibrium states, or sometimes even the motion, of an actual elastic structure.

If we confine our attention to the behaviour of the system in a reasonable vicinity of its initial state, where generally the elastic deformations are *small*, but nevertheless *finite*, then our task becomes much easier, as an approximate form of the relation (3.2) may be quite adequate for the insights we seek. However, later on, this limitation in the analysis must be considered when the results are compared with the experimental evidence.

The obvious next step is to expand the total potential energy function V into a Taylor series at some fixed values of the coordinates Q_i^o, corresponding to the initial state. To facilitate somewhat the manipulation of equations, we define an appropriate notation to be used in forming this expansion. Let us, therefore, introduce the summation conventions over repeated indices used in tensor analysis, but not the use of tensors. We shall denote the operator

$$\frac{\partial}{\partial Q_i} \text{ or } \frac{\partial}{\partial q_i} = (\quad),i \quad \text{and} \quad \frac{\partial}{\partial p} = (\quad),p$$

simply by a comma followed by the index (i) behind the quantity to be differentiated partially with respect to Q_i or q_i.

Finally, we define the total potential energy function of the system under consideration as

$$V = V\,(Q_i,\ P) \quad ; \ i = 1,\ldots,n. \tag{3.3}$$

This function is assumed to satisfy all the continuity requirements in the adjacent coordinate space. It is to be noted that the initial state here is *not* necessarily identical with the *undeformed* state of the structure.

Denoting the change in V, by small v the Taylor series takes initially the form

$$v = V,_i q_i + \frac{1}{2!}V,_{ij}q_iq_j + \frac{1}{3!}V,_{ijk}q_iq_jq_k + \ldots + p(V,_p + V,_{pi}q_i + \frac{1}{2!}V,_{pij}\,q_iq_j + \ldots \tag{3.4}$$

no sum on p.

sum on $i,j,\ldots = 1,2,\ \ldots,n$

The last equation then defines the total potential energy function in the vicinity of the initial or an observed state in which the system may be in *motion* or in *equilibrium*. First of all we note that different terms have different orders of magnitude while the changes q_i remain small. At this stage we also recall from elementary mechanics that equilibrium at a point q_i, $i = 1,\ldots,n$ is conditioned by a stationary value (not necessarily a minimum!) of the total potential energy function. Since the coordinate changes q_i in Eq.(3.4) are *independent* this means that all the coeficients $V,_i$; $i = 1,\ldots,n$ vanish, *if the initial* state is also one of *statical equilibrium*. Adjacent equilibrium positions within the accuracy of Eq.(3.4) are found if the condition for the stationary value of the total potential energy function is established in this vicinity and this is accomplished by differentiating partially this function with respect to all the coordinates of deformation (keeping p fixed) and equating these derivatives to zero. Thus we find using Eq. (3.4), the (n) conditions:

$$\frac{\partial v}{\partial q_\alpha} = v,_\alpha = V,_\alpha + V,_{\alpha i}q_i + \frac{1}{2!}V,_{\alpha ij}q_iq_j + \ldots + p(V,_{p\alpha} + V,_{p\alpha i}q_i + \ldots \tag{3.5}$$

sum on $i,j = 1,\ldots,n$

But, putting $p=0$, $q_i=0$, $i=1,\ldots,n$, we recover the equilibrium condition for the initial state

$$V_{,\alpha} = 0 \quad , \quad \alpha = 1,\ldots,n \tag{3.6}$$

so that the adjacent *equilibrium paths* in this case are defined by

$$V_{,\alpha i}q_i + \frac{1}{2!}V_{,\alpha ij}q_iq_j + \ldots + p(V_{,p\alpha} + V_{,p\alpha i}q_i + \ldots) = 0 \tag{3.7}$$

sum on $i,j=1,\ldots,n$
$\alpha=1,\ldots,n$

The equilibrium paths of a locally linear system, Fig.3.1, are obtained by zeroing the differential coefficients $V_{,\alpha ij}$, $V_{,p\alpha i}$ etc. preceding the non-linear terms. Thus

$$V_{,\alpha i}q_i + V_{,p\alpha}p = 0 \quad ; i, \alpha = 1,\ldots,n \tag{3.8}$$

where $[V_{,\alpha i}]$ now represents the stiffnes matrix of linear systems. This solution is possible on the condition that not all the coefficients $V_{,p\alpha}$ are zero. (If all the coefficients $V_{,p\alpha}$ were zero, the stiffness matrix $[V_{,\alpha i}]$ would have to be singular. We shall encounter this case later on, when the stiffness matrix will be recognized to represent actually the

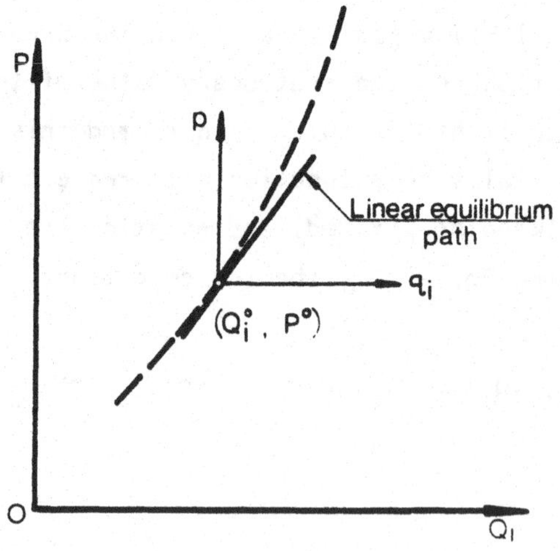

Fig. 3.1.

stability matrix of non-linear systems).

Suppose now we deal with a structure or an elastic system such that all the coefficients

$$V_{,p\alpha} = 0 \quad ; \quad \alpha = 1,\dots,n \tag{3.9}$$

and similarly that;

$$V_{,pp\alpha} = V_{,ppp\alpha} = \dots : 0$$

(These coefficients are not considered in the approximate equation 3.4).

Eq.(3.7) now reduce to the form

$$V_{,\alpha i} q_i = -\frac{1}{2!} V_{,\alpha ij} q_i q_j - V_{,p\alpha i} p q_i - \dots \tag{3.10}$$

We note that the equilibrium equations (3.10) are satisfied when $q_i = 0$, $i=1,\dots,n$ but that p need not be zero. This suggests that the load parameter axis represents an *equilibrium path* in the vicinity of the observed (initial) equilibrium state. This also means that a second equilibrium path exists for values of q_i different from zero, which branches from the same state. When the initial state coincides with the stress-free state of the elastic body in the system (we shall see that for many common structures composed of inextensible bars, the initial state is a stress-free state in this case), the deflected path branches from some *critical* value on the load-parameter axis, which may be reached at no deformations of the elastic body in the system.

As $q_i \to 0$, $i=1,2,\dots,n$, these equations reduce in the limit to the linear form

$$V_{,\alpha i} \left(\frac{q_i}{q_R} \right) = 0 \tag{3.11}$$

where q_R is now a very small reference coordinate. But if q_i are finite and non-zero, then in the observed state, which is now said to be critical,

$$|V_{,\alpha i}(\underset{\sim}{Q}^{o}, P^{o})| = 0 \qquad (3.12)$$

i.e. the determinant of the stiffness matrix in the Eqs.(3.11) vanishes in the initial equilibrium state. Since the initial values Q_k^o of the generalized coordinates of deformation are usually known for the observed state, Eq.(3.12) may be used to determine the critical value of the load parameter P^o. This equation is usually transcendental, so that several or even an unlimited number of *eigen*- values of P may be obtained. Usually the lowest eigenvalue is of practical interest here. The determinant given by Eq.(3.12) is referred to as the *stability determinant* of the system. Once the eigenvalue P^o is known, the linearised equilibrium equations (3.11) can be easily solved. The solutions take the form:

$$q_i = a_i\, q_R + \ldots \qquad (3.13)$$

where a_i are known constants of the system ($a_R = 1$).

3.2. Extention of the Eigenvalue Problem. Solution of the Non-linear Equilibrium Equations.

Let the co-factor of the element $V_{,\alpha\beta}$ in the stability determinant $|V_{,\alpha i}|$ be denoted by the symbol $A_{\alpha\beta}$, where β represents an *arbitrary* column. We multiply the first of the equations (3.10) by $A_{1\beta}$, the second by $A_{2\beta}$, etc. and add all the equations together. The result is

$$(V_{,\alpha 1}A_{\alpha\beta})q_1 + (V_{,\alpha 2}A_{\alpha\beta})q_2 + \ldots + (V_{,\alpha\beta}A_{\alpha\beta})q_\beta + \ldots = -\frac{1}{2!}(V_{,\alpha ij}A_{\alpha\beta})q_i q_j - (V_{,p\alpha i}A_{\alpha\beta})pq_i - \ldots$$

sum on ,i,j = 1,...,n (3.14)

no sum on

But, $(V_{,\alpha\beta}\, A_{\alpha\beta}) = |V_{,\alpha i}| = 0$ represents the development of the stabi i ty determinant by the elements of the βth column. All the other express ns of the sums in the parentheses on the left side of Eq.(3.14) represent expansions of this determinant by *alien co-factors* and are, therefore, identically equal to zero. This means the whole left side of this equation reduces to zero. In this way a *unique** condition is obtained

$$-\frac{1}{2!}(V,_{\alpha ij}\, A_{\alpha\beta})q_i q_j - (V,_{p\alpha i} A_{\alpha\beta})pq_i - \ldots = 0 \qquad (3.15)$$

which defines the *deflected equilibrium path branching from the initial path* represented by the load-parameter axis. We refer to this path as the simple *post-buckling path* of statical equilibrium for non-symmetrical systems. Then, solving for p, the analytical expression of this path is obtained,

$$p = -\frac{1}{2!}\frac{(V,_{\alpha ij}\, A_{\alpha\beta})q_i q_j}{(V,_{p\alpha i}\, A_{\alpha\beta})q_i} - \ldots \qquad (3.16)$$

If now Eq.(3.13) is used, the last result becomes:

$$p = -\frac{1}{2!}\frac{(V,_{\alpha ij}\, A_{\alpha\beta})a_i a_j}{(V,_{p\alpha i}\, A_{\alpha\beta})a_i}\, q_R - \ldots = b_1 q_R + \ldots \qquad (3.17)$$

where b_1 is another known constant of the system. This result is reproduced graphically in Fig. 3.2. This form of equilibrium is verified in Fig. 6.2a and 6.2b where the counterclockwise rotation $(+Q_P)$ of the loaded joint corresponds to the positive coordinate q_R. It will be shown in Section 5 that the rising branch of this path is statically stable while the falling branch is obviously unstable. In practical structures the measurable displacements s_m of the applied loads are usually the more appropriate quantities in describing the equilibrium paths. To show that, in the case considered, the additional displacement Δs_m of an external force F_m from the critical state is a quadratic function of the coordinate *changes*

---

* *The same result is obtained on inserting the column vector on the R.H.S. of Eq.([illegible]) into any column in the stability determinant and expanding this determinant, equated to zero, by the elements of this column.*

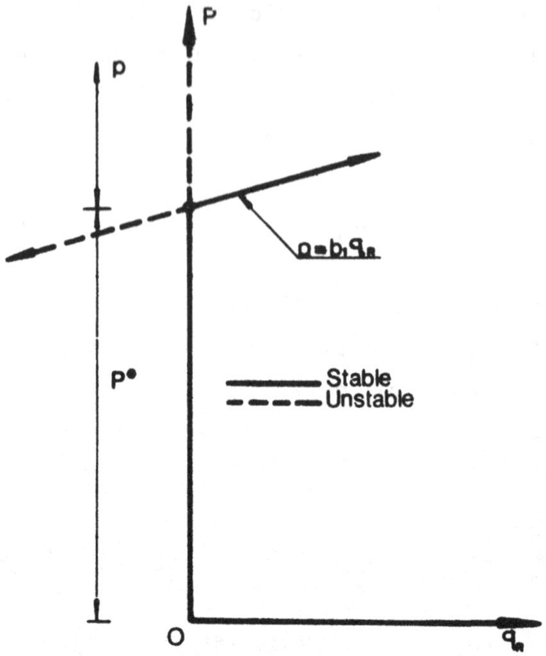

Fig. 3.2.

q_i we expand s_m, F_m and the strain energy U of the system in Taylor series originating in the fundamental equilibrium state (Q_i^o, P^o).

We then form the equilibrium conditions using Eq. (1), i.e.

$$\frac{\partial V}{\partial q_i} = \frac{\partial U}{\partial q_i} - F_m \frac{\partial s_m}{\partial q_i} = 0 \qquad ; \qquad i=1,2,\dots,n \tag{3.18}$$

since now

$$\frac{\partial H}{\partial q_i} = F_m \frac{\partial s_m}{\partial q_i} \tag{3.19}$$

sum on m=1,2,...,M

where M is the number of external loads F_m, these being independent of q_i. The Taylor expansions for U, s_m and F_m may be written:

$$U = U^o + U^o_{,i} q_i + \frac{1}{2!} U^o_{,ij}\, q_i q_j + \dots$$

$$s_m = s_m^o + s_{m,i}^o q_i + \frac{1}{2!} s_{m,ij}^o q_i q_j + \ldots \tag{3.20}$$

$$F_m = F_m^o + F_{m,p}^o p + \ldots$$

sum on $i,j=1,2,\ldots,n$

Here, p is the load-parameter common to all the external loads F_m. Substituting this in Eq.(3.18) we find that equilibrium in the fundamental state is ensured if

$$V_{,\alpha}^o = V_{,\alpha}^o - F_m^o s_{m,\alpha}^o = 0$$

Similarly, using Eq.(3.9)

$$V_{,p\alpha}^o = - F_{m,p}^o s_{m,\alpha}^o = 0 \tag{3.21}$$

$\alpha=1,\ldots,n$

sum on $m=1,\ldots,M$

and the last two conditions are generally satisfied only if $s_{m,\alpha}=0$, and $U_{,\alpha}^o=0$, $\alpha=1,2,\ldots,n$ i.e. the coresponding displacements Δs_m and the strain energy change $\Delta U=U - U^o$ are quadratic functions of the coordinates q_i. Therefore, the change in the corresponding displacement s_m of the load F_m becomes

$$\Delta s_m = s_m - s_m^o = \frac{1}{2!} s_{m,ij}^o q_i q_j + \ldots \tag{3.22}$$

If the expressions (3.20) are substituted into the equilibrium conditions (3.18) these become

$$(U_{,\alpha i}^o - F_m^o s_{m,\alpha i}^c) q_i + \frac{1}{2!} (U_{,\alpha ij}^o - F_m^o s_{m,\alpha ij}^o) q_i q_j - F_{m,p}^o s_{m,\alpha i}^o p q_i + \ldots = 0 \tag{3.23}$$

But, as

$$U_{,\alpha i}^o - F_m^o s_{m,\alpha i}^o = V_{,\alpha i}^o$$

$$U^{o}_{,\alpha ij} - F^{o}_{m}\, s^{o}_{m,\alpha ij} = V^{o}_{,\alpha ij} \tag{3.24}$$

and

$$- F^{o}_{m,p}\, s^{o}_{m,\alpha i} = V^{o}_{,p\alpha i}$$

the original equilibrium equations (3.10)

$$V^{0}_{,\alpha i} q_i + \frac{1}{2!} V^{o}_{,\alpha ij} q_i q_j + V^{o}_{,p\alpha i} p q_i + \ldots = 0 \tag{3.10}$$

$\alpha=1,\ldots,n$

sum on $i,j=1,\ldots,n$

are recovered from (3.23). Therefore, the equilibrium path conditioned by these equations may now be represented in the form

$$p = \pm\, k_{1/2} \sqrt{s - s^{o}} = \pm k_{1/2} \sqrt{s} \tag{3.25}$$

where s^{o} is zero in a system of axially inextensible members, s is a typical displacement and $k_{1/2}$ a constant. Eq. (3.25) is written here for the case when the critical point A is on the p-axis. The graphical representation of this result is shown in Fig. 3.3. This general result is verified experimentally in Figs. 2.2 and 2.3 for the buckling behaviour of two simple non-symmetrical systems. In most applications, such as in the statical analyses of the *buckling behaviour* of plane trusses, the equilibrium equations of the type of Eq. (3.10) can be derived directly from the *conditions of equilibrium* of internal moments and the internal and external forces, as well as from some geometrical requirements.

The solution of these equations is then the same as in the general case considered here. We conclude this discussion by noting that *all non-symmetrical systems are generally unstable* in the critical state A, since, at a constant load parameter P^{o}, a loss of statical stability will occur.

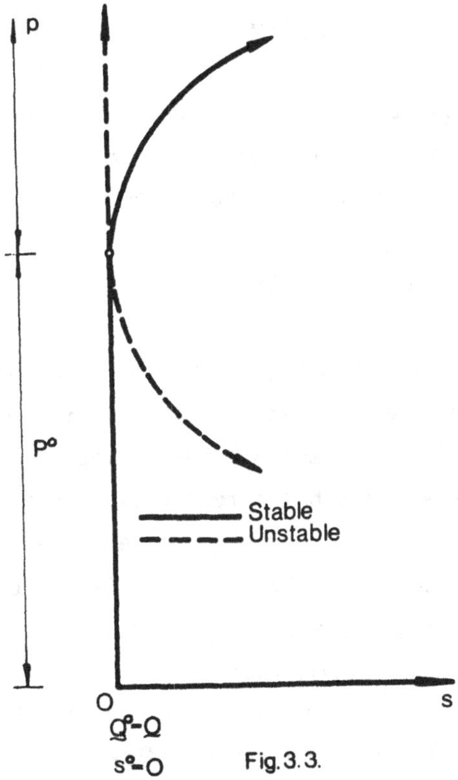

Fig. 3.3.

4. THE EQUILIBRIUM PATHS OF COMPLETELY SYMMETRICAL ELASTIC SYSTEMS

Suppose that we require that the change in the total potential energy function v be a *symmetrical* function in the vicinity of the fundamental state A of the system with respect to the coordinate changes q_i i.e. that

$$v(+q_i) = v(-q_i) \text{ for all } i=1,\dots,n \tag{4.1}$$

It is easily seen that space elements or *entire systems* composed of *pinned* inextensible members readily meet this requirement and that the generalized coordinates are in this case adequately represented by the *end-rotations* of these members. This holds regardles, whether the system represents a plane or a space structure. The series (3.4) in section 3 now simplifies considerably. Some coefficients in this series vanish on account of symmetry, e.g. $V,_i=0$, $V,_{ij}=0$, $V,_{ijk}=0$, $V,_{pi}=0$, etc. for all odd powers in q_i. Since $V,_i=0$, a symmetrical system is necessarily in equilibrium in the fundamental state A. Moreower, since also $V,_{ij}=0$ for $i\neq j$, the stability matrix is now diagonal and the coordinates q_i are *orthogonal*. The expression for the change v in the total potential energy now reduces to:

$$v = \frac{1}{2!} V,_{ii}q_i^2 + \frac{3}{4!} V,_{iijj}q_i^2 q_j^2 + \frac{1}{4!} V,_{iiii}q_i^4 + \dots + p(\frac{1}{2!}V,_{pii}q_i^2 + \dots$$

$$i\neq j;\ \text{sum on } i,j:1,\dots,n \tag{4.1a}$$

Since, also the coefficients $V,_{pi}= V,_{ppi}=\dots=0$, this implies indirectly that the load-parameter axis again represents an equilibrium path.

Thus, v becomes an even function of the coordinates q_i. It is readily shown that the equilibrium equations reduce to

$$v,_\alpha= V,_{\alpha\alpha}q_\alpha + \frac{1}{2!}V,_{\alpha\alpha ii}\, q_\alpha q_i^2 + \frac{1}{3!}V,_{\alpha\alpha\alpha\alpha}q_\alpha^3 + \dots + p(V,_{p\alpha\alpha}q_\alpha + \dots = 0$$

$$\text{no sum on } \alpha=1,\dots,n;\ \text{sum on } i=1,\dots,n;\ i\neq\alpha \tag{4.2}$$

The trivial solutions of these equations is given by .

$$q_\alpha = 0 \quad ; \quad \alpha=1,\dots,n$$

and

$$p \neq 0$$

which means that the *ordinate* or the *p-axis* represents an equilibrium path at *no deformations* of the system. The *non-trivial* solutions are obtained on dividing the n equations 4.2 by q_α ; $\alpha=1,\dots,n$ resulting in the conditions

$$V_{,\alpha\alpha} + \frac{1}{2!} V_{,\alpha ii}\, q_i^2 + \frac{1}{3!} V_{,\alpha\alpha\alpha\alpha}\, q_\alpha^2 + p(V_{,p\alpha\alpha} + \dots \qquad (4.3)$$

no sum on $\alpha=1,\dots,n$; sum on $i=1,\dots,n$; $i \neq \alpha$

Puting $q_i=0$; $i=1,\dots,\alpha,\dots,n$, we get the important condition that for $p=0$

$$V^o_{,\alpha\alpha} = 0 \; ; \quad \alpha=1,2,\dots,n \qquad (4.4)$$

The coefficients* $V^o_{,\alpha\alpha}$ represent the *trace* of the *stability matrix* and they are referred to usually as the *stability coefficients*. These coefficients, therefore, vanish in this case *in the critical state*. The coordinates of deformation q_i are then said to be *critical*.** They are also *orthogonal*.*** It is of interest to verify the condition $V^o_{,\alpha\alpha}=0$ for a particular type of symmetrical systems. Consider, therefore, a structure consisting of *pin-jointed* inextensible, elastic bars, which belong to this category. We assume, for simplicity, the bars to be prismatic. When the critical condition is reached a certain number of critically stressed

---

* *The zero superscript is added for clarity marking the critical state.*

** *See op.cit. Section 1-3*

*** *Loc.cit.*

bars begin to contract flexurally and thereby elastic energy is stored within the system. We construct the expression for the total potential energy change of the system by considering the energy changes for the individual members, then sum up these changes to obtain the *total change*. After buckling the *components* P_i of the *external* forces remain *parallel* to their original directions, Fig.4.1a. Let the angle of body rotation of the buckling bar be (ψ) (positive counterclockwise) and let the force in the direction of the rotated chord of the bar be $P^o + \delta P$ where P^o is the *critical* force equal to the Euler load, so that $P^o = P_{iE} = P_{i(crit)}$. The *flexural* contraction of the bar (shortening of its chord) is denoted by (e). Then, the change in the total potential energy of this bar, with respect to the initial position, is

$$v_i = U_i - P_i [L_i - (L_i - e_i) \cos \psi_i] \qquad (4.5)$$

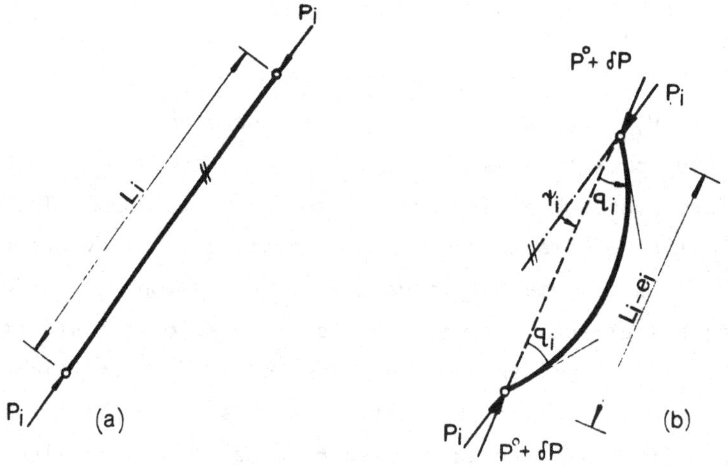

Fig. 4.1.

where U_i is the strain energy change with regard to the *undeformed* state of the bar. Note that in the rotated position (b) the bar is *no longer in statical equilibrium* but in some state of motion, since equilibrium cannc now be maintained by the external forces P_i, (the equilibrium forces $P^o + \delta P$ having been removed).

It can be shown* for this case, that the strain energy U_i of an inextensible bar is given by,

$$U_i = P_{iE} L_i (1 - \cos q_i - \frac{e_i}{L_i}) \tag{4.6}$$

where q_i is the end-rotation of this bar.

For small changes in q_i from the *underformed* state, this reduces to

$$\Delta U_i = U_i = P_{iE} L_i (\frac{1}{2} q_i^2 - \frac{e_i}{L_i} + \ldots) \tag{4.7}$$

It can equally be shown** that in the first approximation

$$\frac{e_i}{L_i} = \frac{1}{4} q_i^2 + \ldots \tag{4.8}$$

Substituting above we get

$$U_i = \Delta U_i = P_{iE} e_i + \ldots \tag{4.9}$$

Expanding now the expression 4.5 for small values of the angle (ψ_i) gives

$$v_i = \Delta U_i - P_i [e_i + \frac{1}{2} (L_i - e_i) \psi_i^2]$$

The body rotation (ψ_i) is of the same order of magnitude in a truss-type elastic system as the flexural contraction parameters (e_i/L_i), so that,

---

* *See: S.J.Britvec, "The Stability of Elastic Systems", Ch. II, Section 2-6, p.150*

** *Op.cit. Ch. II, Section 2-14, Eq.(14.5) and Appendix I*

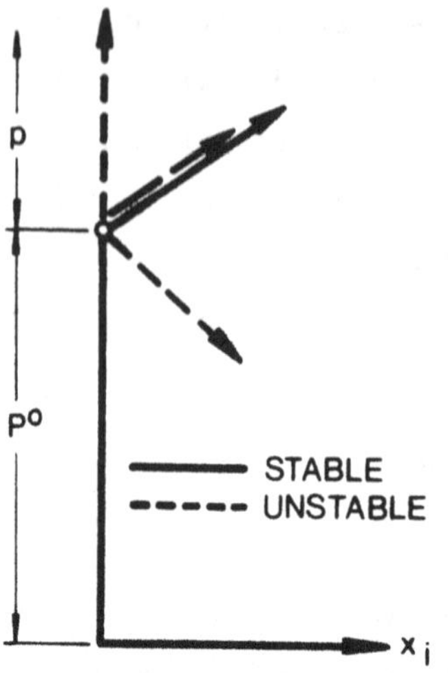

Fig. 4.2.

to the order of (e_i/L_i), Eq. (4.10) may be written, on substituting from Eq. (4.9),

$$v_i = \frac{1}{4} L_i (P_{iE} - P_i)\, q_i^2 + o(q_i^4) \qquad (4.11)$$

from which the *stability coefficients* $V,_{ii}$ can be recovered i.e.

$$V,_{ii} = \frac{1}{4} L_i (P_{iE} - P_i) \quad ; \quad i:1,2,\ldots,n \qquad (4.12)$$

where n is the number of the buckling bars in this case, The condition $V^o,_{ii}=0$ in this system is, therefore, synonimous with the statement that

$$P_{iE} - P_i^o = 0 \quad ; \quad i=1,2,\ldots,n \qquad (4.13)$$

i.e. that the critical axial forces have the Euler values, *which is a well known result.*

The branching paths for this case are shown in Figs. 4.2 and 6.3 and these may be found by solving the Eqs.(4.3).($x_i=q_i^2$ in this case.) We note this result applies regardless of the position of the buckling bars within the system or whether these bars are *statically redundant or not.* Therefore, these non-linear equilibrium paths are typical of both *iso-static and hyper-static reticulated symmetrical systems.*

It is pointed out that in complex structural systems, such as that shown in Fig. 1.4, *accumulation of several critical states*, may lead to very complex *coupled* forms of buckling. This applies to symmetrical as well as non-symmetrical systems. For details the reader is referred to Refs. 10 and 12 (op.cit.).Numerical procedures for the computation of the post-buckling equilibrium paths of reticulated shells of the symmetrical type are treated in Ref.19.

5. STABILITY OF THE EQUILIBRIUM PATHS NEAR THE SIMPLE BRANCHING POINTS

To detect the instability of a non-symmetrical system the relative local change of the total potential energy must be tested for its sign. It is well known from mechanics* that such a change must be negative if the system is unstable in the observed state. To this end expression (3.4) for v is expanded into a *second* Taylor series at a neighbouring point B near the critical state A according to the formula

$$\Delta v = v,_i h_i + \frac{1}{2!} v,_{ij} h_i h_j + \ldots = (h_1 \ldots h_n) \begin{pmatrix} v,_{11} \ldots v,_{1n} \\ \vdots \quad\quad \vdots \\ v,_{n1} \ldots v,_{nn} \end{pmatrix} \begin{pmatrix} h_1 \\ \vdots \\ h_n \end{pmatrix} = \underset{\sim}{h}^T [v,_{ij}] \underset{\sim}{h} + \ldots \tag{5.1}$$

where $v,_i = 0$, because of equilibrium in the observed state which is also assumed to be non-critical. The matrix $[v,_{ij}]$ is non-singular in this case, and the coordinate changes h_i in the physical coordinates q_i are also *non-orthogonal*. They represent increments in the coordinates q_i. To test the signs of Δv the quadratic form in h_i is put into the form of complete squares using the orthogonal coordinates z_i. Then

$$\Delta v = \frac{1}{2!} v,_{ij} h_i h_j = \frac{1}{2!} (\lambda_1 z_1^2 + \lambda_2 z_2^2 + \ldots + \lambda_n z_n^2) + \ldots \tag{5.2}$$

Where $\lambda_1, \lambda_2, \ldots, \lambda_n$ are the local stability coefficients, i.e. the eigenvalues of the *loacl* stability matrix $[v,_{ij}]$ at $\underset{\sim}{B}$.

If $\underset{\sim}{h}$ and $\underset{\sim}{z}$ are supposed to be related linearly through a matrix B, i.e.

$$\underset{\sim}{h} = \underset{\sim}{B}\, \underset{\sim}{z} \tag{5.3a}$$

---

* *See op.cit. Ch. I. The unstable state is followed by an onset of motion so that buckling is really dynamic.*

and if the column vectors of B are determined from the condition

$$[v,_{ij}]\underset{\sim}{b}_j = \underset{\sim}{I}\,\lambda_j\,\underset{\sim}{b}_j \quad \text{or} \quad \{[v,_{ij}] - \underset{\sim}{I}\,\lambda_j\}\underset{\sim}{b}_j = \underset{\sim}{0} \tag{5.3b}$$

where λ_j ; $j=1,2,\ldots,n$ are constants of proportionality, then these can be found from the condition:

$$\begin{vmatrix} v,_{11}-\lambda_j & v,_{12} & \cdot & \cdot & v,_{1n} \\ v,_{21} & v,_{22}-\lambda_j & \cdot & \cdot & v,_{2n} \\ \cdot & \cdot & \cdot & \cdot & \\ \cdot & \cdot & \cdot & \cdot & \\ v,_{n1} & v,_{n2} & \cdot & \cdot & v,_{nn}-\lambda_j \end{vmatrix} = 0 \tag{5.3c}$$

For each λ_j the corresponding column vector of $\underset{\sim}{B}$ can then be found from Eq. (5.3b) where it can also be shown* that $\underset{\sim}{B}^{-1} = \underset{\sim}{B}^T$. Then, using the relation (5.3a) in Eq. (5.1), the result of Eq. (5.2) can easily be derived. The coordinates z_i are now measured along the *orthogonal* eigenvectors (column vectors) $\underset{\sim}{b}_j$ which rotate as B moves on the path, as it is shown in the Appendix.

The coefficients $v,_{\alpha\beta}$ may be computed directly from Eq. (3.4). Thus

$$v,_{\alpha\beta} = V,_{\alpha\beta} + V,_{\alpha\beta i}q_i + V,_{p\alpha\beta}p + \ldots \tag{5.4}$$

In the local orthogonal coordinates at point A, which is critical, this determinant becomes:**

---

* For a detailed elucidation of this and other points in the Eigenvalue theory, the reader is referred to Ref.12, ch. I, section 1-2.

** [illegible] down to a zero or a non-zero element.

$$\begin{vmatrix} V_{,11}-\lambda_j & & \cdot & 0 \\ 0 & V_{,22}-\lambda_j & \cdot & 0 \\ \cdot & \cdot & \cdot & \cdot \\ 0 & 0 & \cdot & V_{,nn}-\lambda_j \end{vmatrix} = 0 \qquad (5.4a)$$

where $V_{,11}$, etc. are the stability coefficients at A. Since A is the first critical point, one of these coefficients, say $V_{,11}$, must be zero. Therefore, also λ_1 is zero at A. If B is only slightly removed from A, then λ_1 at B must be a small quantity, while the local coordinates z_i at A and B are referred to the coordinate system of the local orthogonal eigenvectors, which is rotated as the point moves along an equilibrium path from A to B (see the Appendix). Since all the other stability coefficients $V_{,22},\dots,V_{,nn}$ at A are large and positive (because the system is stable by definition before A is reached!) they differ at B from those at A only by small amounts so that the sign of Δv and the stability of the system at B is settled by the small stability coefficient (first eigenvalue) λ_1^B at point B. Thus

$$\lambda_1^B = \lambda_1^A + \Delta\lambda = \Delta\lambda \quad \text{since } \lambda_1^A = 0 \qquad (5.5)$$

To compute λ_1^B the expressions (5.4) are substituted into (5.3) using the condition that also $V_{,11} = 0$. Thus

$$\begin{vmatrix} 0 + V_{,11i}\, q_i + V_{,p11}\, p - \lambda_1^B & V_{,12} + V_{,12i} q_i + V_{,p12}\, p \\ V_{,21} + V_{,21i}\, q_i + V_{,p21}\, p & V_{,22} + V_{,22i}\, q_i + V_{,p22}\, p - \lambda_1^B \\ V_{,31} + V_{,31i}\, q_i + V_{,p31}\, p & V_{,32} + V_{,32i} q_i + V_{,p32}\, p \end{vmatrix} = 0$$

sum on $i = 1,2,\dots,n$ (5.6)

If $A_{\alpha\beta}$ is the cofactor of the element $V_{,\alpha\beta}$ in the stability determinant at A, then, multiplyng the first row of the last determinant by $A_{1\beta}$, where β is an arbitrary column, the second by $A_{2\beta}$, etc. and adding these rows to the first row, the value of the determinant is not changed, Eq. (5.6a).

If then, the first column is multiplied by q_1, the second by q_2, etc. we get

$$\begin{vmatrix} (V_{,\alpha 1}A_{\alpha\beta})q_1+[(V_{,\alpha 1j}A_{\alpha\beta})q_j+(V_{,p\alpha 1}A_{\alpha\beta})p]q_1-A_{1\beta}q_1\lambda_1^B & (V_{,\alpha 2}A_{\alpha\beta})q_2+[(V_{,\alpha 2j}A_{\alpha\beta})q_j+(V_{,p\alpha 2}A_{\alpha\beta})p]q_2-A_{2\beta}q_2\lambda_1^B & \cdot \\ V_{,21}q_1+(V_{,21j}q_j+V_{,p21}p)q_1 & V_{,22}q_2+(V_{,22j}q_j+V_{,p22}p)q_2-q_2\lambda_1^B & \cdot \\ V_{,31}q_1+(V_{,31j}q_j+V_{,p31}p)q_1 & V_{,32}q_2+(V_{,32j}q_j+V_{,p32}p)q_2 & \cdot \\ & & \cdot \end{vmatrix} = 0$$

sum on $\alpha, j = 1,\ldots,n$ (5.6a)

Now the sums $(V_{,\alpha 1}A_{\alpha\beta})$, $(V_{,\alpha 2}A_{\alpha\beta})$, sum on α, are zero because they either represent the expansion of the stability determinant by alien cofactors or the stability determinant at A itself, which is also zero. If next, in the new determinant, the second column is added to the first, and so is the third up to the nth, then the value of the resulting determinant does not change again. Thus,

$$\begin{vmatrix} [(V_{,\alpha ij}A_{\alpha\beta})q_j + (V_{,p\alpha i}A_{\alpha\beta})p]q_i - A_{i\beta}q_i\lambda_1^B & [(V_{,\alpha 2j}A_{\alpha\beta})q_j + (V_{,p\alpha 2}A_{\alpha\beta})p]q_2 - A_{2\beta}q_2\lambda_1^B & \cdot \\ V_{,2i}q_i + (V_{,2ij}q_j+V_{,p2i}p)q_i - q_2\lambda_1^B & V_{,22}q_2 + (V_{,22j}q_j + V_{,p22}p)q_2-q_2\lambda_1^B & \cdot \\ V_{,3i}q_i + (V_{,3ij}q_j+V_{,p3i}p)q_i - q_3\lambda_1^B & V_{,32}q_2+(V_{,32j}q_j + V_{,p32}p)q_2 & \cdot \\ & & \cdot \end{vmatrix} = 0$$

sum on $\alpha, i, j = 1,2,\ldots,n$ (5.6b)

If now the linear terms $V_{,\alpha i}q_i$; $\alpha = 1,2,\ldots n$, sum on $i = 1,2,\ldots,n$, in the first column of the new determinant are replaced by the non-linear terms using the equilibrium equations (3.10), establishing thereby that B is a point *on the equilibrium path*, the following condition for the computation of λ_1^B may be obtained

$$\begin{vmatrix} [(V_{,\alpha ij}A_{\alpha\beta})q_j + (V_{,p\alpha i}A_{\alpha\beta})p]q_i - A_{i\beta}q_i\lambda_1^B & (V_{,\alpha 2j}A_{\alpha\beta})q_j + (V_{,p\alpha 2}A_{\alpha\beta})p - A_{2\alpha}\lambda_1^B & \cdot \\ \frac{1}{2} V_{,2ij}q_iq_j - q_2\lambda_1^B & V_{,22} + V_{,22j}q_j + V_{,p22}\, p - \lambda_1^B & \cdot \\ \frac{1}{2} V_{,3ij}q_iq_j - q_3\lambda_1^B & V_{,32} + V_{,32j}q_j + V_{,p32}\, p & \cdot \\ & & \cdot \end{vmatrix} = 0$$

sum on $\alpha, i, j = 1,2,\ldots,n$ (5.7)

If this determinant is expanded according to the elements of the first column, then to the third order of small quantities the result is

$$[(V,_{\alpha ij}A_{\alpha\beta})q_iq_j+(V,_{p\alpha i}A_{\alpha\beta})pq_i-A_{i\beta}q_i\lambda_1^B]\ A_{11}+o(q_iq_jq_k)=0 \qquad (5.8)$$

sum on $\alpha, i, j, k = 1,2,\ldots,n$

As B moves closer to point A the third order terms become vanishingly small, so that the criterion for the computation of λ_1^B becomes simply

$$(V,_{\alpha ij}A_{\alpha\beta})q_iq_j + (V,_{p\alpha i}A_{\alpha\beta})pq_i - A_{i\beta}q_i\lambda_1^B = 0 \qquad (5.9)$$

Δv changes sign from positive to negative, i.e. the system loses its stability at the stability boundary when $\lambda_1^B = 0$ or, from the last condition,

$$p = \frac{(V,_{\alpha ij}A_{\alpha\beta})q_iq_j}{(V,_{p\alpha i}A_{\alpha\beta})q_i} + \ldots \qquad (5.10)$$

sum on $i, j = 1,2,\ldots,n$

Comparing this result with Eq.(3.16), it follows that also in physical coordinates* the gradient of the stability boundary near the simple branching point A *has twice the value of the gradient to the equilibrium path* at that point. This is of practical interest as this boundary may now be constructed experimentally using the empirical slope of the equilibrium path. This result is shown graphically in Fig. 5.1.

In the case of the symmetrical systems, physical coordinates are mutually *orthogonal*. The stability boundary then depends on the mode of buckling, and it is parabolic in form, such that the curvature of this boundary at the critical point is *three times the curvature of the path.*

---

* *Compare this with the result in orthogonal coordinates op.cit. p. 57*

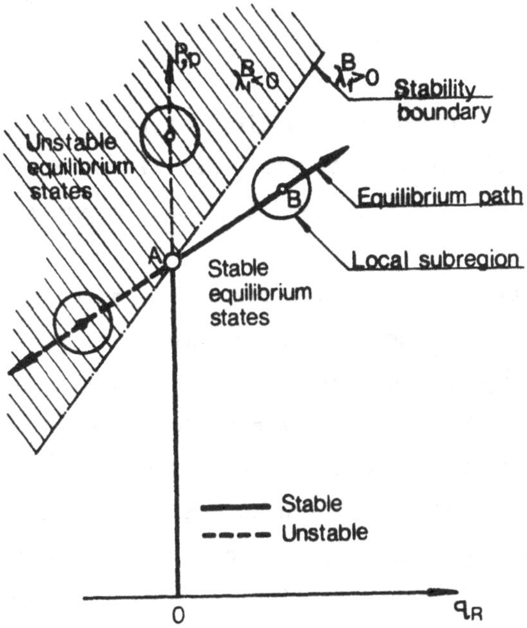

Fig. 5.1.

This case is treated in (ref.12, p.61) and the result is represented graphically in Figs. 6.3a and b.

6. IMPERFECTION - SENSITIVITY OF ELASTIC SYSTEMS

So far we have discussed the elastic systems under *perfect* conditions, which were entirely defined by the generalized coordinates of deformation Q_i (or their increments q_i) and the load parameter P or its change p.

Imperfect conditions of a system may be characterized by several parameters. These may be of a geometrical nature or they may be related to the material or the physical properties of the system. These imperfection parameters may affect the shape of the equilibrium paths of the equivalent perfect system *qualitatively*, so that the original paths are altered beyond recognition. We consider here only the source of *geometrical* imperfections by introducing a *single* parameter α, which is then linked in some way to the initially *slightly* distorted configuration of the elastic system, as it is related to the initial or perfect configuration.

This initial distortion may be envisaged in such a way that all the characteristic geometrical quantities necessary to define the distorted configuration are proportional to α. In this way the distorted configuration is then *quantitatively* represented by α. Obviously, that an unlimited number of such initially distorted shapes may be imagined, each depending on a different relationship to the parameter α. Essentially, the totality of these initially distorted configurations is, in the first instance, reflected in some *initial values of the generalized coordinates* q_i^o, which define *any* distorted, physically possible shape of the system. Therefore, if the initial, imperfect configuration is to depend on a single parameter α, these initial values q_i^o in the generalized coordinates must be essentially proportional to α. It may be argued that additional coordinates could be introduced to define possible initially distorted shapes of the given perfect system, so that the set of coordinates of the imperfect system may be larger than that of the equivalent perfect system. Yet, such an increased set would generally contain as the subset, the coordinates of the equivalent perfect system. We admit, therefore, the possibility that slightly different *initially* distorted shapes would result in *slightly* different equilibrium paths, if in each case such a shape were represented by a single parameter α, but that quantitively this single parameter would, at least approximately, characterize the *shape of the resulting equilibrium paths of the totality of the equivalent slightly imperfect systems.* Also, geometrical imperfections *do not affect the conservative character* of this system.

The total potential energy function V, in this case, depends also on the parameter α. Any change in V will also depend on α and *on the partial derivatives of* V *with respect to* α. Obviously, these derivates will be in turn conditioned by a *single* imperfect shape at a time, so that the different geometrical properties of these shapes will be accounted for in this way.

We define the total potential energy function V^I of an imperfect system, having a single load parameter P, and a single imperfection parameter α, by

$$V^I = V^I\ (\underset{\sim}{Q}, P, \alpha) \tag{6.1}$$

where $\underset{\sim}{Q}$ is the vector representing all the coordinates $Q_i = Q_i^o + q_i$. As before, we next expand Eq. (6.1) in a Taylor series about the branching point $[\underset{\sim}{Q}^o, P^o]$ of the perfect system, treating α as another coordinate in this expansion. This gives:

$$\begin{aligned}\Delta V^I = v^I = {} & V_{,p}p + V_{,\alpha}\alpha + V_{,i}q_i + \frac{1}{2!}\left(V_{,pp}p^2 + V_{,\alpha\alpha}\alpha^2 + V_{,ij}q_iq_j + 2V_{,p\alpha}p\alpha + 2V_{,pi}p\,q_i + 2V_{,\alpha i}\alpha q_i\right) \\ & + \frac{1}{3!}\left(V_{,ppp}\,p^3 + V_{,\alpha\alpha\alpha}\alpha^3 + 3V_{,pp\alpha}p^2\alpha + 3V_{,p\alpha\alpha}\,p\alpha^2 + 3V_{,ppi}p^2q_i + 3V_{,pij}p\,q_iq_j + V_{,ijk}q_iq_jq_k\right) + \dots\end{aligned} \tag{6.2a}$$

We omit hereafter the superscript (I), marking the imperfect system, and establish the non-trivial equilibrium conditions as before by differentiating this series partially with respect to the coordinates q_γ, $\gamma = 1,2,\dots,n$, and by equating the result of this differentiation to zero. Thus

$$V,_{\gamma} = V,_{\gamma} + V,_{\gamma i}q_i + V,_{p\gamma}p + V,_{\alpha\gamma}\,\alpha + \frac{1}{2!}\,(V,_{pp\alpha}\;p^2 + 2V,_{p\gamma i}\;pq_i + V,_{\gamma ij}\;q_iq_j$$
$$+ V,_{\alpha\alpha\gamma}\alpha^2 + 2V,_{\alpha\gamma i}\;\alpha q_i + 2V,_{p\alpha\gamma}\;p\alpha) + \ldots = 0$$

sum on $i,j = 1,2,\ldots,n$ (6.2)

no sum on $\gamma = 1,2,\ldots,n$

On putting $\alpha = 0$, the equilibrium paths of the perfect system are recovered. Therefore, the differential coefficients $V,_{\gamma}$ $V,_{p\gamma}$ $V,_{pp\gamma}$ etc., evaluated at the fundamental point ($\underset{\sim}{Q} = \underset{\sim}{Q}^o$, $P = P^o$, $\alpha = 0$) vanish, since the origin of the Taylor series is the *same for both systems.*

The equilibrium paths of an equivalent slightly imperfect system are then given approximately by

$$V,_{\gamma i}q_i = -\frac{1}{2}V,_{\gamma ij}\;q_iq_j - V,_{p\gamma i}\;p\,q_i - V,_{\alpha\gamma}\,\alpha - \frac{1}{2!}V,_{\alpha\alpha\gamma}\,\alpha^2 - V,_{\alpha\gamma i}\;\alpha q_i - V,_{p\alpha\gamma}\;p\alpha - \ldots;\; \gamma,i,j, = 1,2,\ldots,n$$

(6.3)

To solve these equations we proceed as in the case of the perfect system. We multiply the first of the equations (6.3) by $A_{1\beta}$, the second by $A_{2\beta}$, etc. and add all the equations, where generally $A_{\gamma\beta}$ is the cofactor of the element $V,_{\gamma\beta}$ in the local stability determinant. Thus

$$(V,_{\gamma 1}A_{\gamma\beta})q_1 + (V,_{\gamma 2}A_{\gamma\beta})q_2 + \ldots + (V,_{\gamma\beta}A_{\gamma\beta})q_\beta + \ldots = -\frac{1}{2!}(V,_{\gamma ij}A_{\gamma\beta})q_iq_j - (V,_{p\gamma i}A_{\gamma\beta})pq_i - (V,_{\alpha\gamma}A_{\gamma\beta})\,\alpha + o(\alpha^2, \alpha q_i, p\alpha)$$

(6.4)

sum on $\gamma,i,j = 1,2,\ldots,n$

no sum on $\beta = 1,2,\ldots,n$

Since $|V,_{ji}|$ is the stability determinant of the equivalent perfect system, this determinant vanishes according to Eq.(3.12) in section 3. The left sides of the last equations represent either the expansions of this determinant by alien cofactors, and these vanish identically, or the expansion of the stability determinant itself at A (which is the simple branching point of the equivalent perfect system) and such a determinant is zero. Therefore, the left sides of the last equations are always zero. We obtain,

$$p(V_{,p\gamma i}A_{\gamma\beta})q_i = -(V_{,\alpha\gamma}A_{\gamma\beta})\alpha - \frac{1}{2!}(V_{,\gamma ij}A_{\gamma\beta})q_i q_j - \dots \qquad (6.5)$$

sum on $\gamma, i, j = 1,2,\dots,n$

β = either 1 or 2,...,or n

This result establishes the form of the equilibrium paths of a geometrically slightly imperfect system, i.e. then

$$p = -\frac{(V_{,\alpha\gamma}A_{\gamma\beta}) + \frac{1}{2!}(V_{,\gamma ij}A_{\gamma\beta})q_i q_j}{(V_{,p\gamma i}A_{\gamma\beta})\, q_i} \qquad (6.6)$$

It is to be noted that in Eqs.(6.3) p has a singularity at q=0. But, if the point on the path is sufficently removed from this singularity the approximate solution (3.13) for perfect systems holds since the effect of a small parameter α becomes less significant. Therefore, approximately Eq, (3.13) applies also in this case.

$$q_i = a_i q_R + \dots \qquad (3.13)\text{ bis}$$

Therefore,

$$p = -\frac{(V_{,\alpha\gamma}A_{\gamma\beta}) + \frac{1}{2!}(V_{,\gamma ij}A_{\gamma\beta})a_i a_j q_R^2}{(V_{,p\gamma i}A_{\gamma\beta})a_i\, q_R} + \dots \qquad (6.6a)$$

The maximum or minimum values p^* and q^* are found on differentiating p with respect to q_R and equating this slope to zero. Thus

$$p^* = -\frac{(V_{,\gamma ij}A_{\gamma\beta})q_i^* q_j^*}{(V_{,p\gamma i}A_{\gamma\beta})q_i^*} = -\frac{(V_{,\gamma ij}A_{\gamma\beta})a_i a_j}{(V_{,p\gamma i}A_{\gamma\beta})a_i}\, q_R^* \qquad (6.7)$$

Comparing this result with the equation (5.10) of the stability boundary, we see that maxima or minima, where a stable path becomes unstable or vice versa, *lie precisely on the stability boundary for this system.* Substituting now in Eq. (6.6a) we find

$$q_R^* = \pm\sqrt{\frac{2(V_{,\alpha\gamma}A_{\gamma\beta})}{(V_{,\gamma ij}A_{\gamma\beta}a_i a_j)}}\quad\sqrt{\alpha} \tag{6.8}$$

and

$$p^* = \pm\sqrt{\frac{2(V_{,\gamma ij}A_{\gamma\beta}a_i a_j)(V_{,\alpha\gamma}A_{\gamma\beta})}{(V_{,p\gamma i}A_{\gamma\beta}a_i)}}\quad\sqrt{\alpha} \tag{6.9}$$

sum on $\gamma, i, j = 1,2,\ldots,n$

Therefore, the peaks of the imperfect paths are removed from A by distances proportional to $\sqrt{\alpha}$, and this result agrees with earlier evidence.* The paths are reproduced in Fig. 6.1.

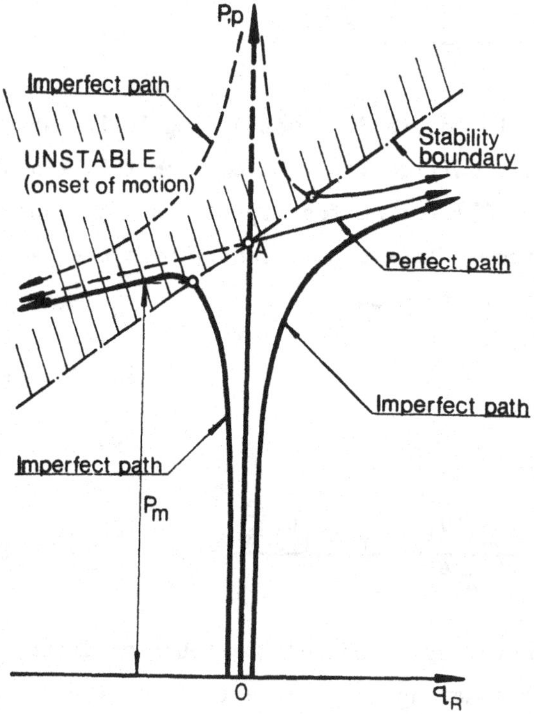

Fig. 6.1.

* *Refs.4,5,12,17,18. These curves are verified experimentally on a portal model-frame in Figs.6.2a and b. If in Eq.(6.9) no real root exist, then the curves are as shown in Fig.6.2a.*

Similarly it may be shown for symmetrical systems,* that the imperfect paths are as indicated in Fig. 6.3a and b.

In this case the maximum or minimum is removed from A in such a way that p^* is proportional to $\alpha^{2/3}$ and q_R^* to $\alpha^{1/3}$, where α is again a small geometrical imperfecion parameter. The peak load P_m, occurs again only in the case of unstable paths.

Of most interest for practical applications is *the imperfection sensitivity* of a particular system. A measure of this sensitivity is the maximum load P_m which is on one side conditioned by the imperfections of the system, while, on the other side, it largely depends *on the shape of the post-buckling paths* of the equivalent perfect system. Therefore, if the post-buckling paths are flat, *as this is usually the case for symmetrical systems*, then also the peak loads P_m are likely to occur on the level of the critical load P^o_c while in highly unstable systems, with decreasing parabolic paths with convexities towards the load parameter axis, the reduction of the peak load may be considerable. In some cases this reduction is of the order of 50% or more.

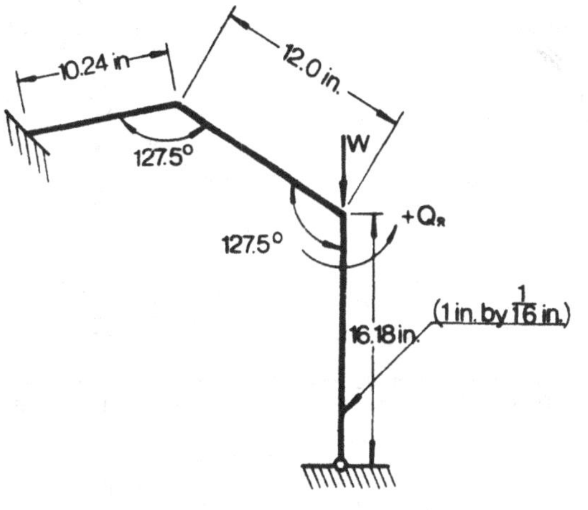

Fig 6.2. Simple portal frame. (Refs. 12 and 18)**

* *See op.cit. p.61 and 112*

** *Model frame used for testing the equilibrium forms in Fig.6.1(Ref.18). Experimental results are reproduced in Figs.6.2a and 6.2b.*

To assess the imperfection sensitivity of a structural system, the power law, which determines the relationship between the peak load P_m, Fig.6.3b, and the imperfection parameter α of the system, must be resonably well known. The 2/3 power law applies to symmetrical systems and the 1/2 power law to non-symmetrical structural systems. For a more general group of structural systems this law may be stated in the form*

$$P_m = P^o \, (1 - C\,\alpha^n) \qquad (6.10)$$

where n may be determined experimentally or otherwise for a particular kind of system. P^o is the critical load obtained from the eigen-value theory and C is a constant. Then, for non-symmetrical systems, n=1/2 and for symmetrical systems n=2/3.

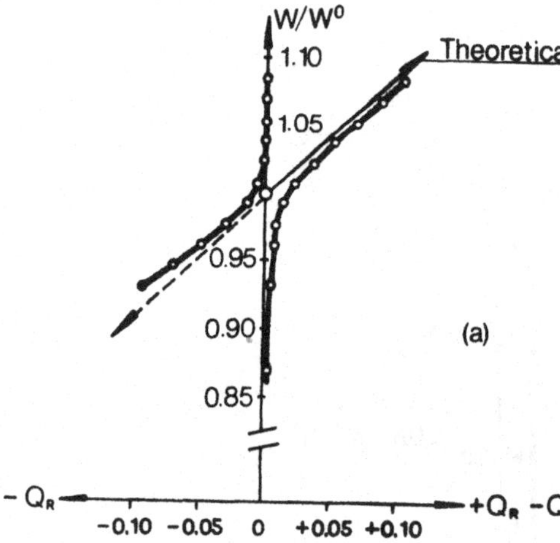

Fig. 6.2.(a)**
(W° is the critical load)

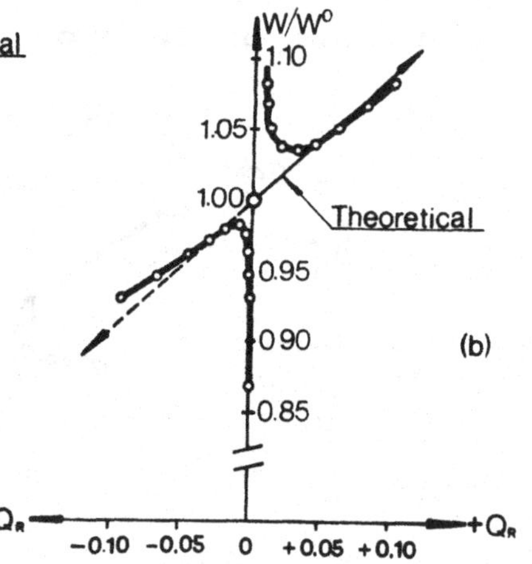

Fig. 6.2.(b)

---

* *Author, op.cit.*

** *See footnote on previous page (Ref.18)*

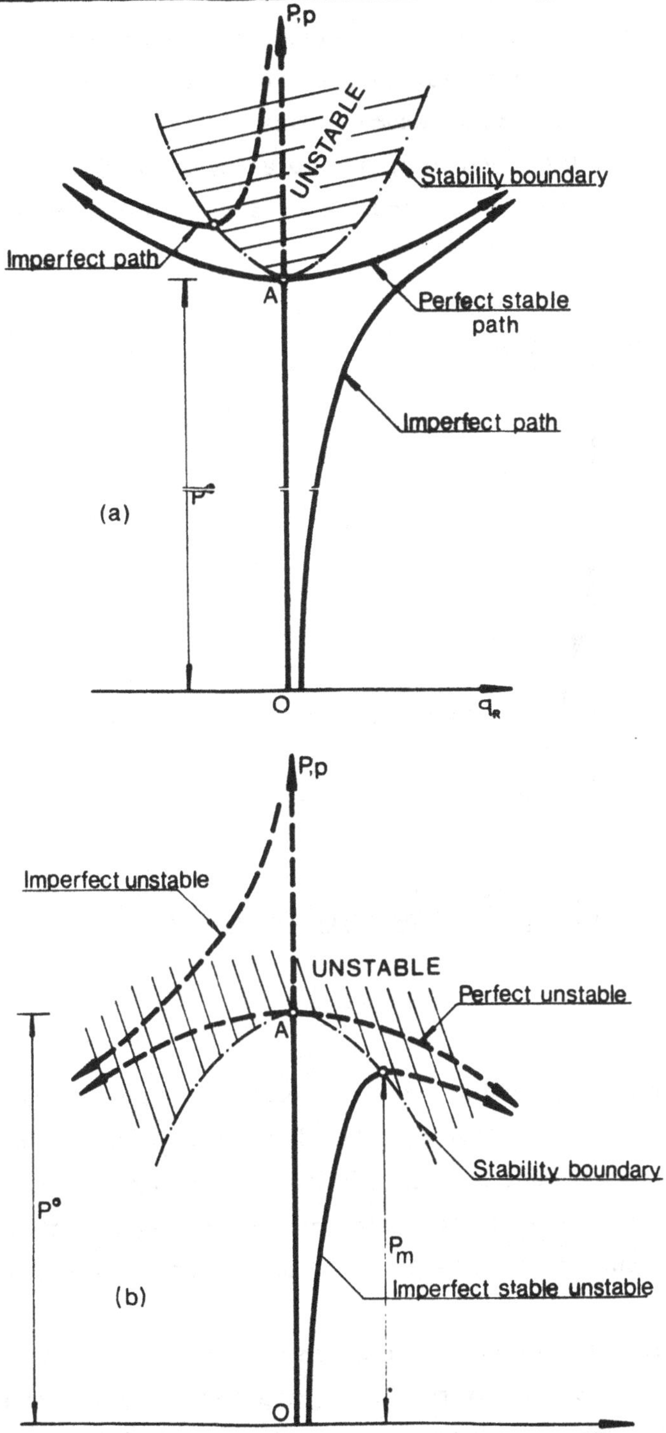
P,p
UNSTABLE
Stability boundary
Imperfect path
A
Perfect stable
path
Imperfect path
P°
(a)
O
q_R
P,p
Imperfect unstable
UNSTABLE
Perfect unstable
A
Stability boundary
P°
P_m
(b)
Imperfect stable unstable
O
q_R

Fig. 6.3.

When C is large the load reductions are large for small imperfections. Then, even at small imperfections, a large scatter in peak loads is possible, at which the system becomes unstable.

For small C the conditions are more favourable, because a relatively small scatter in peak loads occurs over a wider range of practical imperfections. See Figs. 6.4a and 6.4b.

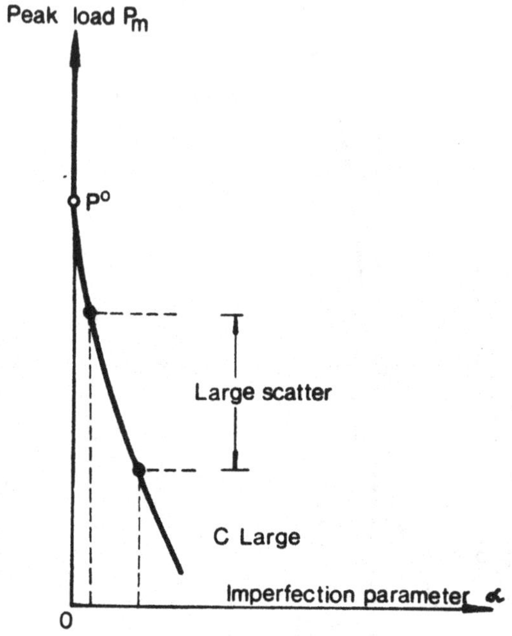

Fig.6.4.a

For pinned or quasi-pinned structures, such as plane and space trusses or reticulated shells with flat post-buckling paths, C may be small, so that *scatter of maximum loads over a wide range of imperfections is small.* Therefore, instability of the system over a wide range of these imperfections may *not* occur, as long as the load-parameter remains beneath a certain value.

Many practical systems, such as shells or elastically jointed frameworks, may belong to an intermediate group for which the dependence on geometrical imperfections may be more complex than 1/2 or 2/3 power law. Also, the scatter of peak loads P_m may be severe for small imperfections

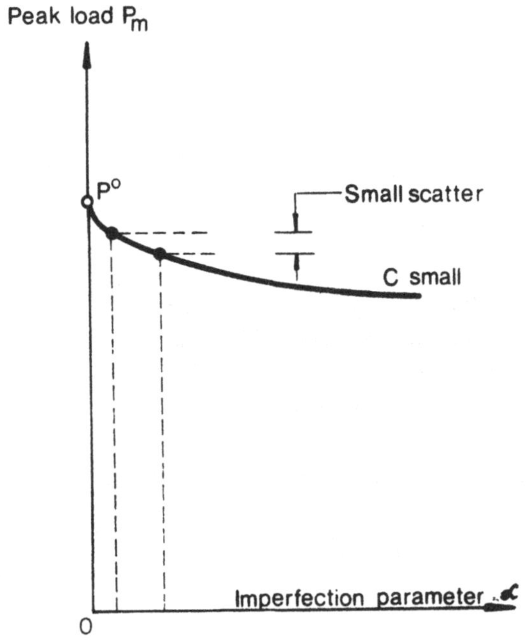

Fig.6.4.b

and less severe for larger imperfections. In that case the imperfection-sensitivity curve has a *bottom or plateau* as indicated in Fig. 6.5. One such structural type is a sandwich cylindrical shell or ring subjected to temperature changes, when different layers of the shell have different thermal expansion properties; Ref. 9.

Structurally useful systems are those for which C is small or those for which this plateau exists, since this ensures that safe maximum design loads may be prescribed with confidence over a practically unlimited range of initial imperfections.

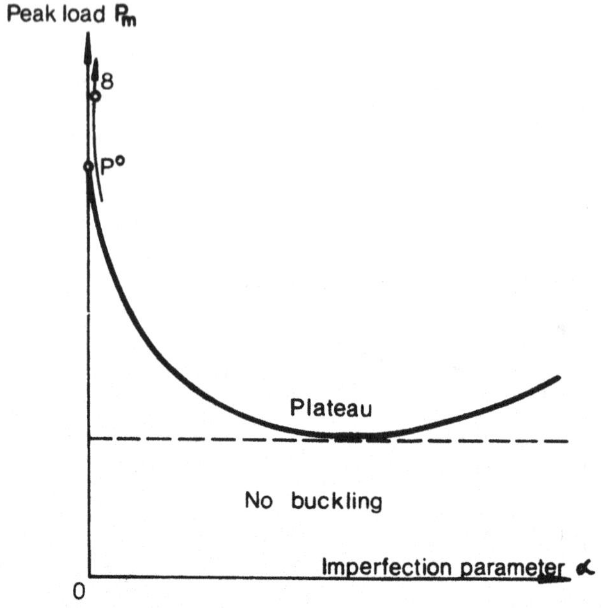
Peak load P_m
8
P°
Plateau
No buckling
Imperfection parameter α
0

Fig. 6.5.

7. APPENDIX

It remains to be shown that actually the local orthogonal coordinates z_i at any point B on an equilibrium path may be thought of as the coordinates in a local coordinate system defined by the local *eigenvectors* $\underset{\sim}{b}_j$, and that this coordinate system rotates as point B moves along this path.

For this purpose consider a system with three coordinates h_1, h_2 and h_3. Then according to Eq. (5.3a) these depend on the coordinates z_1, z_2 and z_3 through the relations

$$\begin{aligned} h_1 &= b_{11}\, z_1 + b_{12}\, z_2 + b_{13}\, z_3 \\ h_2 &= b_{21}\, z_1 + b_{22}\, z_2 + b_{23}\, z_3 \\ h_3 &= b_{31}\, z_1 + b_{32}\, z_2 + b_{33}\, z_3 \end{aligned} \qquad (1.)$$

or $\underset{\sim}{h} = B\,\underset{\sim}{z}$

and these may be written

$$\begin{pmatrix} h_1 \\ h_2 \\ h_3 \end{pmatrix} = \begin{pmatrix} b_{11} \\ b_{21} \\ b_{31} \end{pmatrix} z_1 + \begin{pmatrix} b_{12} \\ b_{22} \\ b_{32} \end{pmatrix} z_2 + \begin{pmatrix} b_{13} \\ b_{23} \\ b_{33} \end{pmatrix} z_3 \qquad (2.)$$

or $\underset{\sim}{h} = \underset{\sim}{b}_1 z_1 + \underset{\sim}{b}_2 z_2 + \underset{\sim}{b}_3 z_3$

If now, b_{1j}, b_{2j}, b_{3j}; $j=1,2,3$ are conceived as the Cartesian components of the vector $\underset{\sim}{b}_j$. Then

$$\underset{\sim}{b}_j = b_{1j}\,\underset{\sim}{i}_1 + b_{2j}\,\underset{\sim}{i}_2 + b_{3j}\,\underset{\sim}{i}_3 \qquad j = 1,2,3 \qquad (3.)$$

where $\underset{\sim}{i}_1, \underset{\sim}{i}_2, \underset{\sim}{i}_3$ are unit vectors.

If now we form the scalar product between $\underset{\sim}{b}_k$ and $\underset{\sim}{b}_j$ for $k \neq j$ we get

$$\underset{\sim}{b}_k \underset{\sim}{b}_j = b_{1k} b_{1j} + b_{2k} b_{2j} + b_{3k} b_{3j} = \underset{\sim}{b}_k^T \underset{\sim}{b}_j$$

This product is zero if the *orthogonality condition* (5.2e) *holds*. Therefore,

$$\underset{\sim}{b}_k \underset{\sim}{b}_j = \underset{\sim}{b}_k^T \underset{\sim}{b}_j = 0 \qquad \text{for } k \neq j$$

means that the two vectors are orthogonal. This means that the three vectors $\underset{\sim}{b}_1$, $\underset{\sim}{b}_2$ and $\underset{\sim}{b}_3$ define a local coordinate system at a particular point (in this case at point B). But, then, the vector on the left side of Eq.(1.) must be equally conceived as having the Cartesian components h_1, h_2 and h_3. For now

$$\begin{aligned}
\underset{\sim}{h} = h_1 \underset{\sim}{i}_1 + h_2 \underset{\sim}{i}_2 + h_3 \underset{\sim}{i}_3 = \underset{\sim}{b}_1 z_1 + \underset{\sim}{b}_2 z_2 + \underset{\sim}{b}_3 z_3 = \\
(b_{11} \underset{\sim}{i}_1 + b_{21} \underset{\sim}{i}_2 + b_{31} \underset{\sim}{i}_3)\, z_1 + \\
(b_{12} \underset{\sim}{i}_1 + b_{22} \underset{\sim}{i}_2 + b_{32} \underset{\sim}{i}_3)\, z_3 + \\
(b_{13} \underset{\sim}{i}_1 + b_{23} \underset{\sim}{i}_2 + b_{33} \underset{\sim}{i}_3)\, z_3 = \\
= (b_{11} z_1 + b_{12} z_2 + b_{13} z_3)\, \underset{\sim}{i}_1 + \\
(b_{21} z_1 + b_{22} z_2 + b_{23} z_3)\, \underset{\sim}{i}_2 + \\
(b_{31} z_1 + b_{32} z_2 + b_{33} z_3)\, \underset{\sim}{i}_3
\end{aligned} \qquad (4.)$$

Comparing now the Cartesian components on both sides of the Eq.(4.) the relations (1.) are recovered. This means that, indeed,

$$\underset{\sim}{h} = \underset{\sim}{b}_1 z_1 + \underset{\sim}{b}_2 z_2 + \underset{\sim}{b}_3 z_3$$

is a vector and that the coordinates z_1, z_2 and z_3 are orthogonal and measured along the axis of the coordinate system defined by the local eigenvectors b_j, $j=1,2,3$ at B. But we have already seen that the condition for the orthogonality of these vectors is the *diagonal* stability matrix at this point (in this case at point A), Eq. (5.4a). Therefore, if at a neighbouring point B on the equilibrium path this matrix is *not* orthogonal, i.e. its non-diagonal elements $V^B_{,ij} = V^A_{,ij} + \Delta V_{,ij}$ are *not* zero when $\Delta V_{,ij}$ is expressed by the coordinate changes h_i and p, both of which statisfy the equation of this path at B, then the local *orthogonal* coordinate system at B with the new local orthogonal coordinates z_i is *not* co-axial with the local orthogonal coordinate system at A, *but rotated in relation to it*. This is then tantamount to the statement that the local system of orthogonal axis *rotates* as this point moves along the path.

REFERENCES

1. Britvec, S.J., *The Post-Buckling Behaviour of Frames*, Dissertation, Cambridge University, 1960.

2. Britvec, S.J., Overal Stability of Pin-jointed Frameworks after the Onset of Elastic Buckling, *Ingen.-Archiv*, Vol.32/6, pp.443-452, 1963.

3. Britvec, S.J., Elastic Buckling of Pin-jointed Frames, *Int.J.Mech.Sci.* Vol.5 pp.447-461, 1963.

4. Britvec, S.J. and Chilver, A.H., Elastic Buckling of Rigidly-jointed Plane Frames, *J.Eng.Mech.Div.ASCE*, Vol.89, pp.217-225, 1963.

5. Britvec, S.J., The Theory of Elastica in the Non-linear Behaviour of Plane Frameworks, *Int.J.Mech.Sci.*, Vol.7, pp.661-684, 1965.

6. Britvec, S.J., Some Aspects of the Elastic Stability of Pin-jointed Space Frameworks and Reticulated Shells, *Space Structures*, pp.471-484, John Wiley & Sons, Inc., N.Y., 1967.

7. Britvec, S.J., Stability Critera for Completely Symmetrical Discrete Elastic Systems, *S.R.C.C. Report*, No. 79 (NASA), Pittsburgh, 1968.

8. Britvec, S.J., Notes on Hyperstatic Symmetrical Discrete Elastic Systems and Stability, *S.R.C.C. Report*, No.87 (NASA), Pittsburgh, 1969.

9. Britvec, S.J., Sur le flambage thermique des anneaux et coques cylindriques précontraints, *Journal de Mécanique*, Vol.5, No.4.Déc., 1966.

10. Britvec, S.J., *A Treatise on the Stability of Elastic Systems*, Habilitationsschrift, University of Stuttgart. National Science Foundation, [illegible],Pr.No GK [illegible], University of Pittsburgh, [illegible] August 1972.

11. Britvec, S.J. and Yu, M.T., *Buckling of a Hyperstatic Segment of a Reticulated Cylindrical Shell*, National Science Foundation, Washington, D.C., *Report*, Pr. No. GK 2879, University of Pittsburgh, January,1972.

12. Britvec, S.J., *The Stability of Elastic Systems*, Pergamon Press, New York, N.Y., 1973.

13. Bryan, G.H., On the Stability of Elastic Systems, *Proc.Cambridge Phil. Soc.*, Vol.6, pp.199, 1888.

14. Chilver, A.H. and Britvec, S.J., The Plastic Buckling of Aluminium Columns, *Proceedings*, Symposium on Aluminium in Structural Engineering, Paper 4, pp.1-22, London, 1963.

15. Chwalla E., Die Stabilität lotrecht belasteter Rechteckrahmen, *Der Bauingenier*, Springer-Verlag, Berlin, January,1938.

16. Koiter, W.T., Elastic Stability and Post-Buckling Behaviour, *Nonlinear Problems*, edited by R.E.Langer, The University of Wisconsin Press, 1963.

17. Roorda, J., Stability of Structures with Small Imperfections, *J.Eng. Mech.Div.* ASCE, Vol.91, pp.87-106, 1965.

18. Godley, M.H.R. and Chilver, A.H., Elastic Post-Buckling Behaviour of Unbraced Plane Frames, *Int.J.Mech.Sci.*, Vol.9, pp.323-333, 1967.

19. Britvec, S.J., and Nardini, D., Some Aspects of the Nonlinear Elastic Behaviour and Instability of Reticulated Shell-Type Systems, *Developments in Theoretical and Applied Mechanics, Proceedings*, Eighth SECTAM, Vol.8, April, 1976.

GENERALIZED MEASURES FOR LARGE DEFORMATIONS

B.R. SETH
Birla Institute of Technology
Mesra, Ranchi, India.

1. Introduction

In engineering materials, even in the largest purely elastic deformations, the strain is small even though the displacement gradients and rotations may be large. In such cases there is very little gained in altering classical linear strain-stress tensor constitutive equations. In 1935, a paper by B.R. Seth[1] renewed interest in such problems, and then followed a host of papers by prominent workers in the field of continuum mechanics. A number of them wanted to consider incompressible materials like rubber and produced complicated constitutive equations, for which many adhoc approximations had to be made to get an agreement with experimental results. But with the advent of soft alloys and high polymers the problem of large deformations, involving severe strains, has again put the constitutive equations in the melting pot. Even though all branches of science have adopted Generalized Measures for problems both at the micro and macro levels, the engineer has been reluctant to change his concepts. He does not want to go beyond the Cauchy and Hencky measures. It should also be appreciated that high speed computation of engineering problems can give reliable results only if the order of strain measure is prescribed in advance. Thus arises the need of adopting generalized measures to simplify constitutive equations and to get better agreement with experimental results.

The concept of Generalized Measure has now been adopted by a number of workers. It incorporates more of the non-linear behaviour of a material in the definition of a strain and less of the non-linear behaviour in the constitutive relation between the strain energy density and the strain. Recently Blatz, Tschoegl, Sharda[2] and their co-workers at California Institute of Technology have shown that large deformations of rubberlike materials in which the response co-efficients may vary as much as 500 percent can be explained by taking one or two terms in the strain energy function if Seth's Generalized Measure is used. In England, Ogden[3] has adopted the same technique. Again, Fung[4] has used an exponential measure for rheological problems, and Hill[5] has shown that constitutive inequalities for elastic and plastic deformations depend on the strain measure used. In a recent lecture Jean Mandel[6] has advocated the use of generalized measure in plasticity and visco-elasticity. Hsu, Davies and Royles[7] have made a computer programme for an uni-dimensional generalized measure. Other workers include Narasimhan, Knoshaug, Bakshi, Purushothama, Hulsurker, Borah[8]. We have used it in a number of recent research papers[9].

2. Measures of Deformation

Measures of deformation can be referred to a material system (X^α), called Lagrangian coordinates or to a spatial system (x^α), called Eulerian coordinates. The system x^α is independent of the choice of X^α. In the first we can designate a particle X^α and in the second a place x^α. In mechanics of large deformations the spatial coordinates play an important role. The classical equations of motion and the boundary conditions are formulated in this system; in the other they become very complex. Hence it is essential that the deformation and the stresses be also referred to the spatial coordinates; otherwise, a paradox will arise. For small deformations the two representations become the same, and hence it becomes immaterial which system is employed. In the material system the finite strain tensor, called the Green – St. Venant tensor, may be represented by E_{ij}. In the spatial system it is called the Almansi-Hamel tensor and can be represented by e_{ij}. Both these tensors are symmetric. In cartesian coordinates they differ only by the sign of the second order terms. In fact in terms of the displacement u_i they are

$$2\,e_{ij} = u_{i,j} + u_{j,i} - u_{\alpha,i}\,u_{\alpha,j} \tag{2.1}$$

$$2\,E_{ij} = u_{i,j} + u_{j,i} + u_{\alpha,i}\,u_{\alpha,j}\ . \tag{2.2}$$

In general we have

$$x^i = x^i(X^L, t)\ ,\ X^L = X^L(x^i, t) \tag{2.3}$$

$$dx^i = x^i{}_{,L}\ dX^L\ ,\ dX^L = X^L{}_{,i}\ dx^i$$

$$ds^2 = c_{LM}\ dX^L\, dX^M\ ,\ dS^2 = c_{ij}\ dx^i\, dx^j \tag{2.4}$$

where

$$C_{LM} = g_{ij}\ x^i, L\ x^j, M\ ,\quad c_{ij} = G_{LM}\ X^L,i\ \ X^M,j$$

We easily see that

$$2\,E^L_M = C^L_M - \delta^L_M\ ,\ 2e_{ij} = \delta^i_j - c^i_j\ . \tag{2.5}$$

We have another measure, called Hencky's logarithmic measure, which is widely used in plastic deformations. In terms of the measures e and E it is

$$H = \frac{1}{2}\ \log(1 + 2E)\ ,\ h = -\frac{1}{2}\ \log(1 - 2e)\ . \tag{2.6}$$

The deviators of h and H represent changes of shape without change of volume. In the one-dimensional case this is known to be equal to $\log(1/1_0)$, 1 and 1_0 being the deformed and undeformed lengths respectively.

The two description e and E are not generally equal to each other unless their principal axes are the same or there is a pure strain unaccompanied by rotation.

Besides these three classical measures others have been in use from time to time, but no systematic attempt has been made to generalize them. On the other hand the constitutive equations have received almost all the attention and have been sometimes made too complex to be of any experimental use. Modern technology has to deal with all types of large deformations. So we do need measures whose order can be fixed a priori so that they can simplify the constitutive equations, are better amenable to high speed computation and hence can be expected to give better agreement with experimental results[10].

Cayley-Hamilton theorem enables us to express any measure ϵ_{ij} in terms of another measure e_{ij} by the relation–

$$\epsilon_{ij} = F_0\, \delta_{ij} + F_1\, e_{ij} + F_2\, e_{i\alpha}\, e_{\alpha j}\,, \tag{2.7}$$

F's being functions of the three invariants of e_{ij}. Thus we can express any generalized measure in terms of e, E or H.

When an ordinary measure is found unsuitable it is customary in all branches of science to use weighted measures. Only the engineer has been found shy of adopting this practice, and those interested in mechanics have not come to his help. Take the uni-dimensional case and notice how the known measures can be deduced from the elementary Cauchy measure by the introduction of a weight-function. We have–

$$e = \int_{l_0}^{l} \frac{dl}{l_0} = \frac{l - l_0}{l} \qquad \text{(Cauchy)}\,.$$

The choice of the weight-function should be obviously the stretch ratio (l_0/l). Thus the first weighted-measure is

$$e_1 = \int_{l_0}^{l} \frac{l_0}{l} \cdot \frac{dl}{l_0} = \log\left(\frac{l}{l_0}\right) \qquad \text{(Hencky)}\,.$$

The second is

$$e_2 = \int_{l_0}^{l} \left(\frac{l_0}{l}\right)^2 \frac{dl}{l_0} = \frac{l - l_0}{l}\,. \qquad \text{(Approximate Almansi-Hamel, used by Schwainger).}$$

The third is

$$e_3 = \int_{l_0}^{l} \left(\frac{l_0}{l}\right)^3 \frac{dl}{l_0} = \frac{1}{2}\left(1 - \frac{l_0^2}{l^2}\right). \qquad \text{(Almansi-Hamel)}\,.$$

The obvious generalization is

$$e_{n+1} = \int_{l_0}^{l} \left(\frac{l_0}{l}\right)^{n+1} \frac{dl}{l_0} = \frac{1}{n}\left(1 - \frac{l_0^n}{l^n}\right), \quad \text{(Seth)}. \tag{2.8}$$

Changing n into $-n$ gives Green-Saint-Venant type of measures. e_{n+1} contains the known uni-dimensional measures in use.

For the general case the weight-function to be used for principal measures will be $(dS/ds)^n$, so that the principal generalized measures ϵ_{xx}, ϵ_{XX} in terms of e_{xx}, e_{XX} are given by –

$$\epsilon_{xx} = \int_0^{e_{xx}} (1-2e_{xx})^{\frac{1}{2}n+1}\, de_{xx} = \frac{1}{n}[(1-2e_{xx})^{\frac{1}{2}n}] \tag{2.9}$$

$$\epsilon_{XX} = \int_0^{E_{XX}} (2\,E_{XX} - 1)^{\frac{1}{2}n+1}\, dE_{XX} = \frac{1}{n}[\,(1 + 2E_{XX})^{\frac{1}{2}n} - 1]. \tag{2.10}$$

These can be further generalized by taking powers of ϵ_{xx} of ϵ_{XX}. In the principal tensors we can expand the term on the right hand side and use Cayley-Hamilton theorem. We then get forms for the generalized measure given in (2.7). We can also write the generalized measures as

$$\epsilon^i_j = \frac{1}{n}[\,\delta^i_j - (c^{n/2})^i_j\,] = \sum_\alpha M^i_\alpha \frac{1-\lambda^n_\alpha}{n} M^\alpha_j \tag{2.11}$$

$$\epsilon^k_L = \frac{1}{n}[\,(C^{n/2})^k_L - \delta^k_L\,] = \sum_\alpha N^k_\alpha \frac{\lambda^n_\alpha - 1}{n} N^\alpha_L \tag{2.12}$$

c, C being the Almansi and Green tensors and M^i_α . N^k_L being the matrix of the eigen vectors of their respective tensors, and (λ_α) are the eigenvalues of the corresponding tensors. Thus we see that by using the eigenvalues of the Green and Almansi tensors we can readily generalize our measures to suit problems in large deformation.

As an example, the generalized measures for a radial displacement $u = r(1-\beta)$, $\beta = f(r)$; $\beta' = f'(r)$, $\beta = d\beta/dr$, are given by

$$\epsilon_{rr} = \frac{1}{n}[\,1-(r\beta' + \beta)^n\,], \quad \epsilon_{\theta\theta} = \epsilon_{\phi\phi} = \frac{1}{n}(1-\beta^n) \tag{2.13}$$

$$\epsilon_{r\theta}\,,\ \epsilon_{\theta\phi}\,,\ \epsilon_{\phi r} = 0$$

f being an arbitrary function of r.

3. Applications

i) Rubber-like materials and high polymers.

Many workers have treated rubber and rubber-like materials by taking particular forms of the strain-energy functions. In such cases a simplification is introduced by the deformation being very nearly isochoric. With the extensive use of high polymers it is only recently that attention has been paid to the use of generalized measures in place of complex constitutive equations. P.J. Blatz, S.C. Sharda and N.W. Tschoegl[2] have shown that by using Seth's measures of the type (2.8) only the first term in the strain energy function is sufficient to get good agreement with experimental results for deformation up to about 200 per cent strain in simple tension. They have used them for simple compression, pure shear, simple shear and torsion and have found satisfactory results. In like manner T.C. Hsu, S.A. Davies, R. Royles, Y.C.B. Fung, R.W. Ogden, M.N.L. Narasimhan, V.S. Bakshi, B.N. Borah, S. Hulsurker have used them in solid and fluid mechanics.

Blatz, Sharda and Tschoegl have found that for simple tension, the true Lagrangian stress defined as $T_\alpha = \lambda_\alpha\, \partial w/\partial\alpha$, λ_α being a principal stretch ratio, is given by

$$T_\alpha = (2G/n)\,(\lambda^{n} - \lambda^{-\frac{1}{2}n}) \tag{3.1}$$

G and n are material constants. The ordinary stress is given by

$$T = (2G/n)\,(1 - \lambda^{-\frac{3}{2}n})\,, \tag{3.2}$$

which indicates the asymptotic value $(2G/n)$, when λ becomes very large.

They have found that for $n = 1.64$ their results hold good for natural rubber and styrene-butadiene rubber. These will certainly hold good for many high polymers.

ii) Large longitudinal vibration of strings and rods

If u denotes the displacement, and we use the Almansi measure, we get the differential equation of motion as

$$\frac{\partial^2 u}{\partial t^2} = \frac{E}{\rho_0} \frac{\partial^2 u/\partial x^2}{(1 + \partial u/\partial x)^3} , \tag{3.3}$$

which is the equation for long waves. If the generalized measure in (2.8) is used we get

$$\frac{\partial^2 u}{\partial t^2} = \frac{E}{\rho_0} \frac{\partial^2 u/\partial x^2}{(1 + \partial u/\partial x)^{n+1}} \tag{3.4}$$

which is the equation for adiabatic sound waves. Both (3.3) and (3.4) can be solved in the form

$$\frac{\partial u}{\partial t} = f\,[\,x + u-(c + \partial u/\partial t)t\,]\;\; c^2 = E/\rho_0 , \tag{3.5}$$

f being an arbitrary function.

These show that waves cannot be propagated without change of phase[11].

iii) Vertical oscillations of a mass attached to an elastic string

If l, l′ are the stretched and the unstretched lengths in the position of equilibrium, the time t of an oscillation is given by

$$t = (\frac{ml}{E})^{\frac{1}{2}} (\frac{l'}{l})^{\frac{3}{2}} (\sec\theta)^{\frac{1}{2}} E(k, \phi) , \tag{3.6}$$

where $k = \sin\theta = (\alpha - \beta)/\alpha$ is the modulus of the elliptic function. The cnmplete period is given by

$$t = 4(\frac{ml}{E})^{\frac{1}{2}} (\frac{l'}{l})^{\frac{3}{2}} (\sec.\theta)^{\frac{1}{2}} E_0 . \tag{3.7}$$

E_0 being the complete integral of the second kind[12].

Many other applications of the generalized measures in (2.11) and (2.12) have been worked out by B.R. Seth[10]. These include problems in elastic-plastic deformation and creep.

II. Elastic-plastic Deformation

The classical theory of elastic-plastic deformation has to assume incompressibility and a yield condition. The transition theory developed by us in some recent papers does away with both these assumptions. The setting in of plastic deformation is intimately connected with the geometry of deformation. This involves the use of generalized strain measures, and, in particular, the Almansi measure. The corresponding field equations have critical points. The asymptotic solutions at these points give the required results, of which classical ones are only particular cases[13].

III. Creep Deformation

1. Introduction

Creep deformation has been dealt with in two series of lectures delivered at CISM[16]. All the three stages of creep are taken into consideration. No assumptions regarding incompressibility, yield conditions and creep strain laws are made.

2. Sheeting Bending

Sheet bending into circular and equiangular forms can exhibit both creep and creep-plastic effects[17,19,23,24]. We assume that the orthogonal character of the surfaces before and after deformation is preserved. Also the sheet is taken to be long enough to be treated as a two-dimensional problem in plane strain. The tentative values of the displacement components can be taken as –

$$u = x - x' = x - f(\alpha)$$
$$v = y - y' = y - \phi(\beta) \tag{2.1}$$

where dashes refer to the unstrained state and

$$z = x + iy = F(\alpha + i\beta) = F(\xi) \tag{2.2}$$

f, ϕ, F are functions of α, β and ξ respectively, which have to be determined.

We have taken the strained state as the reference-frame work, which is pertinent to our problem. Referred to it the finite components of strain are –

$$e_{xx} = \frac{1}{2}(1 - f'^2\,\alpha_x^2 - \phi'^2\,\beta_x^2)$$
$$e_{yy} = \frac{1}{2}(1 - f'^2\,\alpha_y^2 - \phi'^2\,\beta_y^2)$$
$$e_{xy} = -\frac{1}{2}(f'^2\,\alpha_x\,\alpha_y + \phi'^2\,\beta_x\,\beta_y) \tag{2.3}$$

where $\alpha_x = \partial\alpha/\partial x$, $\beta_x = \partial\beta/\partial x$ etc.

The corresponding principal strain components are :

$$e_{\alpha\alpha} = \frac{1}{2}(1 - F_2 f'^2), \quad e_{\beta\beta} = \frac{1}{2}(1 - F_2\phi'^2) e_{\alpha\beta} = 0, \tag{2.4}$$
$$F_2 = |d\xi/dz|^2.$$

The generalized strain components are [21]

$$e_{\alpha\alpha} = n^{-m}(1 - F_2^{n/2}\,f'^n)^m,$$
$$e_{\beta\beta} = n^{-m}(1 - F_2^{n/2}\,\phi'^n)^m, \quad e_{\alpha\beta} = 0. \tag{2.5}$$

A linear stress-strain law, sufficient for our purpose, is

$$T_{ij} = \lambda e_{ii} + 2\mu e_{ij}, \quad e_{ii} = e_{\alpha\alpha} + e_{\beta\beta}. \tag{2.6}$$

The equations of equilibrium in α,β coordinates are

$$\frac{\partial}{\partial\alpha}\left(\frac{T_{\alpha\alpha}}{h}\right) = T_{\beta\beta}\frac{\partial}{\partial\alpha}\left(\frac{1}{h}\right), \quad \frac{\partial}{\partial\beta}\left(\frac{T_{\beta\beta}}{h}\right) = T_{\alpha\alpha}\frac{\partial}{\partial\beta}\left(\frac{1}{h}\right) \tag{2.7}$$

$$F_2 = h^2.$$

These may be written as

$$\frac{\partial T_{\alpha\alpha}}{\partial\alpha} - \frac{1}{2}(T_{\alpha\alpha} - T_{\beta\beta})\frac{\partial}{\partial\alpha}(\log F_2) = 0$$

$$\frac{\partial T_{\beta\beta}}{\partial\beta} - \frac{1}{2}(T_{\beta\beta} - T_{\alpha\alpha})\frac{\partial}{\partial\beta}(\log F_2) = 0\,. \tag{2.8}$$

Let us put $A = 1 - F_2^{n/2}\ f'^n\,,\ \ B = 1 - F_2^{n/2}\ \phi'^n.$ (2.9)

The equations (2.8) may now be written as

$$\frac{\partial}{\partial\alpha}[\lambda(A^m + B^m) + 2\mu A^m] - \mu(A^m - B^m)\frac{\partial}{\partial\alpha}(\log F_2) = 0$$

$$\frac{\partial}{\partial\beta}[\lambda(A^m + B^m) + 2\mu B^m] - \mu(B^m - A^m)\frac{\partial}{\partial\beta}(\log F_2) = 0\,. \tag{2.10}$$

3. General solution of equation (2.10)

Equation (2.10) may be written as

$$\frac{\partial}{\partial\alpha}[\log\{(A^m - B^m)F_2^{-c/2}\}] + \frac{2-c}{A^m - B^m}\frac{\partial B^m}{\partial\alpha} = 0 \tag{3.1}$$

$$\frac{\partial}{\partial\beta}[\log\{(A^m - B^m)F_2^{-c/2}\}] - \frac{2-c}{A^m - B^m}\frac{\partial A^m}{\partial\beta} = 0$$

$$c = 2\mu / (\lambda + 2\mu)\,. \tag{3.2}$$

When $c \to 0$, these equations give the solution as

$$A^m + B^m + K_0 \quad \text{(a constant)} . \tag{3.3}$$

When $m = 1$, $n = 2$, we get the known solution[22]

$$F^{2/(2-c)} F_2 = K_1 \quad \text{(a constant)} . \tag{3.4}$$

where $F_1 = f'^n + \phi'^n$.

In general equation (3.1) may be written as

$$(A^m - B^m) F_2^{-c/2} \frac{\partial}{\partial \alpha} [\log \{ (A^m - B^m) / F_2^{-c/2} \}]$$

$$+ m(2-c) B^{m-1} F_2^{-1/2c} \frac{\partial B}{\partial \alpha} = 0 \tag{3.5}$$

which gives

$$\frac{\partial}{\partial \alpha} [(A^m - B^m) F_2^{-c/2}] + m(2-c) \phi'^c B^{m-1} (1-B)^{-c/n} \frac{\partial B}{\partial \alpha} = 0 . \tag{3.6}$$

The general solution of (3.1) may therefore be written as –

$$(A^m - B^m) F_2^{-c/2} = K(\beta) - m(2-c) \phi'^c \int B^{m-1} (1-B)^{-c/n} \frac{\partial B}{\partial \alpha} d\alpha . \tag{3.7}$$

In like manner we get from (3.2)

$$(A^m - B^m) F_2^{-c/2} = K(\alpha) + m(2-c) f'^c \int A^{m-1} (1-A)^{-c/n} \frac{\partial A}{\partial \beta} d\beta \tag{3.8}$$

$K(\alpha)$ and $K(\beta)$ being arbitrary functions of α and β respectively.

The particular case of $m = 1$, $n = 2$, gives the result in equation (3.4), both $k(\beta)$ and $K_1(\alpha)$ becoming a constant. This has been used to treat the elastic-plastic bending of a plane sheet into a cyclindrical shape[15].

4. Turning Point Solution of (3.1) and (3.2)

The Jacobian of the deformation transformation (2.1) is

$$f'\phi' \,|\, d\xi/dz \,|^2 \;=\; f'\phi' F_2 \;.$$

Since $F_2 \neq 0$, the transition points can be

$$f' \to 0 \quad \text{and} \quad \phi' \to 0 \;.$$

These belong to the region under extension and contraction respectively. We shall first discuss $f' \to 0$.

In this case the equation (3.2) or (3.8) gives

$$\frac{\partial}{\partial\beta} \left[\log (A^m - B^m)/F_2^{-c/2} \right] = 0 \tag{4.1}$$

which shows that

$$A^m - B^m \sim K_2(\alpha)\, F_2^{c/2} \tag{4.2}$$

$K_2(\alpha)$ being an arbitrary function of α.

Equation (3.1) can be written with the help of (2.5) and (2.6) as

$$\frac{\partial}{\partial\alpha} \left[\log \frac{T_{\alpha\alpha} - T_{\beta\beta}}{F_2^{c/2}} \right] - (2-c) \frac{\partial}{\partial\alpha} \left[\log (A^m - B^m) \right] + \frac{2-c}{A^m - B^m} \frac{\partial A^m}{\partial\alpha} = 0 \;.$$

Or

$$\frac{\partial}{\partial\alpha} \left[\log \frac{(T_{\alpha\alpha} - T_{\beta\beta})^{-(1-c)}}{F_2^{c/2}} \right]$$

$$-\frac{m(2-c)}{A^m - B^m}\, A^m\, f'^{\,n-1}\, [\, f' \frac{\partial}{\partial \alpha} (F_2^{m/2}) + n\, f''\, F_2^{n/2}\,]\,. \tag{4.3}$$

The second gives the asymptotic value for $f' \to 0\ (n > 1)$ as

$$T_{\alpha\alpha} - T_{\beta\beta} \sim K_2'\,(\beta)\, F_2^{\,c/2\,(1-c)}$$

which gives the elastic-plastic solution[15], but is not suitable for creep deformation. In like manner the first equation gives the asymptotic value for $f' \to 0$ as

$$T_{\alpha\alpha} - T_{\beta\beta} \sim K_2(\beta)\, F_2^{\,c/2}\, (A^m - B^m)^{2-c}$$

$$\sim K_2(\beta)\, F_2^{\,c/2}\, (1 - B^m)^{2-c}. \tag{4.4}$$

Equation (4.2) and (4.4) are to be used for creep-plastic effects[23]. For circular bending the transformation is $z = \exp(k\,\xi)$ and for equiangular spiral bending it is $z = \exp[(a + ib)\,\xi]$[25]. If the spirals are given by $\alpha =$ constant, $\beta =$ constant we have

$$r = d_1 \exp(\theta \cot \alpha_1),\quad r = d_2 \exp(\theta \cot \alpha_2) \tag{4.5}$$

where

$$(a^2 + b^2)\alpha = a \log d_1,\quad (a^2 + b^2)\,\beta = -\,b \log d_2$$
$$\cot \alpha_1 = -b/a,\quad \cot \alpha_2 = a/b$$
$$b^2 = F_2 = (a^2 + b^2)^{-1} \exp[\,2(-a\alpha + b\beta)\,]\,. \tag{4.6}$$

If (4.2) and (4.4) are to be consistent, we should take

$$T_{\alpha\alpha} - T_{\beta\beta} = \exp[\,c'(-a\alpha + b\beta)\,]\,[\,1 - \{\,1 - L_n \exp(-na\alpha)\,\}^m\,]^{2-c} \tag{4.7}$$

L_n being a constant of integration and $c'(\beta) \sim \exp(-b\beta)$.

If all the three stages of creep are taken into account, including the one corresponding to Hencky's measure $(n \to 0)$ we get

$$T_{\alpha\alpha} - T_{\beta\beta} = \exp[3c(-a\alpha + b\beta)]\,[1 - \{1 - L_n \exp(-na\alpha)\}^m]^{2-c} \times$$
$$\times\, [1 - \{1 - L_p \exp(-pa\alpha)\}^m]^{2-c}. \qquad (4.8)$$

When $n.p \to 0$ and $m,q \to 1$, we get with the help of (2.6)

$$T_{\alpha\alpha} - T_{\beta\beta} = K_0\, L_n^2\, L_p^2 = -Y \text{ (a constant)} \qquad (4.9)$$

Y may be called the yield stress. K_0 is a suitably chosen constant. Putting the value of $T_{\alpha\alpha} - T_{\beta\beta}$ in (2.8) and integrating, we can get $T_{\alpha\alpha}$.

If only secondary creep is taken into consideration. as is done currently by many workers we get from (4.7) by putting $m = 1$

$$[c + n(2-c)]\,T_{\alpha\alpha} = Y \exp(cb\alpha)\,[\exp\{-(2n + c - cn)a\alpha\}$$
$$-\exp\{-(2n + c - cn)a\alpha_0\}] \qquad (4.10)$$

For $c \to 0$ it becomes $2nT_{\alpha\alpha} = Y[\exp(-2na\alpha) - \exp(-2na\alpha_0)]$. (4.11)

REFERENCES

[1] Seth, B.R., Finite strain in elastic problems. Phil. Trans. Roy.Soc. London. (A) 234, 231, 1935.

[2] Blatz, P.J., Sharda, S.C., Tschoegl, N.W., Proc. Nat. Acad. Sci., 70, 3041, 1973; Strain energy function of rubberlike materials based on a generalized measure of strain. Trans. Soc. Rheo., 18, 145, 1974; Letters Appl.Eng.Sci., 2, 53, 1974.

Blatz, P.J., Chu, B.M., Wayland, H., Trans.Soc.Rheol., 13, 836, 1969.

Chu, B.M., and Blatz, P.J., Ann.Biomed.Eng. 1, 204, 1972.

[3] Ogden, R.W., Proc.Roy.Soc.Lond. A326, 565, 1972.

[4] Fung, Y.C.B., Amer.J.Physiol., 1, 204, 1972.

[5] Hill, R., J.Mech.Phy.Solids, 16, 229, 315, 1968,

[6] Jean Mandel, Int.Cent.Mech.Sci., Udine, Lecture 97, 1, 1971.

[7] Hsu, T.C., Davies, S.R., and Royles, R., J. Basic Eng. 89, 453, 1967; ASME, Paper no. 66 WA/Met. 1.

Hsu, T.C., Jour. Strain Anly., 1, 331, 1966.

[8] Narasimhan, M.N.L. and Knoshaug, R.N., Int.J. Non-linear Mech., 7, 161, 1972.

Narasimhan, M.N.L., Sra, K.S., Int.J. Non-linear Mech., 4, 361, 1969; Ind.J.Pure App.Math. 3, 549, 1971.

Borah, B.N., Ind.J.Pur.App.Math. 2, 335, 1971; 4, 289, 1973

Bakshi, V.S., Arch.Mech.Stos., 21, 649, 1969;

Purushothama, C.M., ZAMM, 45, 401, 1965;

Hulsurker, ZAMM, 46, 431, 1966.

[9] Seth, B.R., Generalized strain and transition concepts for elastic-plastic deformation, creep and relaxation, Proc. XI Int. Cong. of App.Mech., Munich, 383, 1964.

[10] Continuum concepts of measure, Presidential address, Xth Cong. of Theo. and Appl.Mechanics, Madras 1, 1965.
Measure -concept in mechanics, Int.Jour. of non-linear Mechanics, 1, 35, 1966.

[11] Finite longitudinal vibrations. Proc.Ind.Acad.Sci. A25, 151, 1948.

[12] Some recent applications of the theory of finite elastic deformation. "Elasticity", McGraw Hill, N.Y., Toronto and London, 1950, 67.

[13] Elastic plastic transition in shells and tubes under pressure, ZAMM, 43, 345, 1963.

[14] Elastic -plastic transition in torsion, ZAMM, 44, 229, 1964.

[15] Transition theory of sheet-bending. Prik.Mat.Mek. 27, 380, 1963.

[16] Seth, B.R., Transition problems of aelotropic yield and creep rupture, courses and lectures, Int. Centre. Mech. Sci., Udine, Italy, no. 47, 1970, 1–47; no. 149, 1972, 1–28.

[17] Creep rupture, Proc.IUTAM Symp. on 'Creep in Structures', Göthenburg, 167, 1970.

[18] Transition analysis of collapse of thick cylinders, ZAMM, 50, 617, 1970.

[19] Creep transition, continuum mechanics and related problem of analysis, Moscow, 459, 1972.

[20] Yield conditions in plasticity, Arch.Mech. Stosowanej, Warszawa, 24, 769, 1972.

[21] Measure concept in mechanics. Inter.Jour. Non-linear Mech. 1, 35, 1966.

[22] On a functional equation in finite deformation, ZAMM, 42, 391, 1962.

[23] Creep-plastic effects in sheet-bending, ZAMM, 54, 557, 1974.

[24] Creep transition in rotating cylinders. Jour.Math.Phy. Sci., 8.1, 1974.

[25] Selected problems of applied mechanics, Akad. Nauk, 649, 1974 (Moscow).

CONTENTS

Printed in the United States
By Bookmasters